D0571881

FAST FORWARD

BROOKINGS FOCUS BOOKS

Brooking Focus Books feature concise, accessible, and timely assessment of pressing policy issues of interest to a broad audience. Each book includes recommendations for action on the issue discussed.

Also in this series:
Brain Gain: Rethinking U.S. Immigration Policy
by Darrell M. West

A BROOKINGS FOCUS BOOK

FAST FORWARD

ETHICS AND POLITICS IN THE AGE OF GLOBAL WARMING

William Antholis
Strobe Talbott

BROOKINGS INSTITUTION PRESS
Washington, D.C.

ABOUT BROOKINGS

The Brookings Institution is a private nonprofit organization devoted to research, education, and publication on important issues of domestic and foreign policy. Its principal purpose is to bring the highest quality independent research and analysis to bear on current and emerging policy problems. Interpretations or conclusions in Brookings publications should be understood to be solely those of the authors.

1775 Massachusetts Avenue, N.W., Washington, D.C. 20036
www.brookings.edu

Library of Congress Cataloging-in-Publication data

Antholis, William.
Fast forward : ethics and politics in the age of global warming / by William Antholis and Strobe Talbott.
p. cm. — (A Brookings focus book)
Includes bibliographical references and index.
Summary: "Clearly establishes how and why global warming is a major threat and why urgent action is needed, including the history of domestic and global negotiations on global warming and the players who must be involved in finding a solution to climate change to protect future generations"—Provided by publisher.
ISBN 978-0-8157-0469-0 (hardcover : alk. paper)
1. Global warming—Government policy—United States 2. Global environmental change—Government policy—United States 3. Environmental policy—United States. 4. United States—Environmental conditions. I. Talbott, Strobe. II. Title. III. Series.
QC981.8.G56A57 2010
363.738'740561—dc22 2010015429

9 8 7 6 5 4 3 2 1

Printed on acid-free paper

Typeset in Sabon

Composition by Cynthia Stock
Silver Spring, Maryland

Printed by R. R. Donnelley
Harrisonburg, Virginia

For Annika Elizabeth and Kyri Janet Antholis,

Loretta Josephine and Theodore Brooke Talbott

With this book, the Brookings Institution Press launches the Brookings Focus Books series to help inform public debate on major issues in a timely and accessible fashion.

This book draws on the research of all five Brookings research programs: Governance Studies, Economic Studies, Foreign Policy, Metropolitan Policy, and Global Economy and Development. In particular, *Fast Forward* was written with help from the John L. Thornton China Center, the Arms Control and Nonproliferation Initiative, the Brookings-Blum Roundtable, the Climate and Energy Economics Initiative, the Energy Security Initiative, the Growth through Innovation Forum, the Managing Global Insecurity project, the Global Civics project, and our partners at Climate Central in Princeton, New Jersey.

Our research was funded, in part, by the Brookings Mountain West Initiative, a partnership with the University of Nevada, Las Vegas.

CONTENTS

CHAPTER ONE

THE ACCIDENTAL EXPERIMENT

FOR TENS OF THOUSANDS OF YEARS, we and our ancestors have treated the earth as a laboratory in which we have tinkered with the forces of nature. From taming fire and harnessing wind to developing antibiotics, the results have often advanced civilization. Yet for the past two centuries, we have been conducting what could be the most momentous and dangerous of all experiments: warming the globe.

We started the experiment without meaning to, and, until recently, we did not even know it was under way. Now it may be out of control, threatening to ruin our planet as a home for us and countless other creatures.

Avoiding that fate is a test of our humanity. We flatter ourselves with the anthropological designation *Homo sapiens*. The phrase is often translated simply as "man who knows." We have now, belatedly, met that definition: those of us alive today are the first generation to know that we live in the Age of Global Warming. We may also be the last generation to have any chance of doing something about it. Our forebears had the excuse of ignorance. Our descendants will have the excuse of helplessness. We have no excuse.

But the Latin participle *sapiens* means more than just possessing knowledge; it connotes wisdom, common sense, and competence. By that standard, we have a long way to go—and not much time. It is as though we were watching a video on fast forward. There is still some mystery about what is happening and plenty of suspense about how it will turn out. But we cannot just wait and see. We must respond, and our response, too, must be on fast forward.

Climate change is a test of our scientific and entrepreneurial ingenuity. The necessary restructuring of our industries and economy will be possible only if our leaders demonstrate determination, skill, and courage in their policies for their own nations and in cooperation with one another. So climate change is a test of politics as the art of the possible in the face of what we must hope is only a *nearly* impossible problem.

Climate change is also a test of our ethics, the values that underlie our politics. The potential catastrophe the planet faces obliges us to rethink our rights and duties as citizens. More fundamentally, we need to rethink our obligations as members of a sometimes shortsighted, sometimes sapient, potentially endangered species—and to act accordingly.

A PLANETARY FEVER

During the Industrial Revolution, starting in the late eighteenth century, manual labor and draft-animal farming gave way to the manufacture of goods by machines that ran on energy generated from burning coal. About a hundred years later, in 1897 Mark Twain thought it a witty truism to observe that "everybody talks about the weather, but nobody does anything about it."[1] Yet factories in Europe and North America, including in Twain's adopted home state of Connecticut, were emitting carbon dioxide and other gases in quantities that would shift the balance between the absorption and reflection of solar energy in a way that would risk overheating the planet.

Shortly after Twain died in 1910, temperatures started to creep upward. At first, the warming effects of carbon dioxide were too small to be identified as a trend. Then, around 1970, scientists began to close ranks around the suspicion that something new and worrisome was happening. In 1988 the World Meteorological Organization and the United Nations Environment Program established the Intergovernmental Panel on Climate Change (IPCC). Based on input from more than a thousand of the world's leading meteorologists, geologists, oceanographers, and physicists, the panel concluded that in the course of the twentieth century, the average temperature of the earth's surface had increased 1.3° Fahrenheit from the average in the nineteenth century.

That may not sound like a lot, given the day-to-day fluctuations in weather, not to mention season to season or even year to year. In the winter of 2009–10, for example, three snowstorms, one in December and two in February, dumped a total of four and a half feet of snow on Washington, D.C. The U.S. government shut down for four days; Maryland, Virginia, and Delaware declared a state of emergency; and Senator James Inhofe of Oklahoma mocked the idea of global warming by building an igloo near the U.S. Capitol. In fact, the blizzards neither proved nor disproved the reality of climate change. Neither did Katrina, the Category 3 hurricane that devastated New Orleans in 2005.

The reason for alarm is in the pattern that scientists have discerned over time. A century ago, the numbers of record hot days and record cold days were about the same, whereas in the past decade, there have been about twice as many record highs, and the frequency and severity of storms have increased. And while there have been spikes at both the hot and cold ends of the spectrum, the net effect has been a rise in the average surface temperature of the planet.[2]

Warming since the nineteenth century has initiated the melting of the polar ice caps and the rapid retreat of major glaciers, such as those in the Rockies and the Andes. That phenomenon has begun to deprive the earth of the dual cooling function that ice performs by chilling its surroundings and forming a reflective shield that bounces heat rays back into space.

A few degrees' increase in global average temperature can have a significant impact. Twenty thousand years ago—when it was about 9°F colder than it is today—the ice covering present-day Canada and much of the northern United States was more than a mile thick.[3] As the earth gradually warmed over the following millennia, the combination of temperate climate and stable sea levels was conducive to the enrichment of soil and the growth of fish populations along the coasts. These conditions—a result of natural global warming—opened a new chapter in human history: the Neolithic Revolution, about 10,000 years ago, when nomadic hunter-gatherers settled down on the eastern coast of the Mediterranean and formed stationary communities.

The IPCC has concluded that what is happening now—a sudden and rapid change in weather patterns—is *not* natural; rather, it is largely anthropogenic, a consequence of human activity. The panel also believes that the effects of rising temperatures to date are a likely prelude to more menacing developments in the decades to come.

So far the IPCC has issued four multivolume reports. The last came out in 2007.[4] Each assessment represents an updated consensus, and each update has been more alarming than its predecessor about how fast the planet seems to be warming and more certain about human activity being the cause.

As TEMPERATURES RISE, the danger is analogous to a fever in the human body. We feel healthy at 98.6°F, not so good at 99.5°, and lousy at 101°; if we get up to 105° or so we are likely to be taken to the hospital. So where, on a thermometer that registers the earth's temperature, should there be a marker indicating that a fever is not just uncomfortable and unhealthy but life-threatening? As they contemplate this question, scientists have settled on 3.6°F above average temperatures about a century ago, before they began to rise as a result of the Industrial Revolution. Since the global average has already risen 1.3°F, that means we have only 2.3°F to go before we hit 3.6°F.[5]

That now-canonical figure—3.6°—sounds suspiciously precise, especially in Fahrenheit. It suggests that we can predict within a tenth of a degree when the situation might become catastrophic. On the Celsius or (centigrade) scale, which most of the world uses, the equivalent of 3.6°F is 2°C, which sounds less like a tipping point and more like a focal point on the spectrum of global warming as the disruptive effects grow in frequency, multiplicity, severity, and unpredictability.[6]

The hydrological factor in climate change is another example of questions that are still under debate. Two-thirds of the earth's surface is water. Warming causes evaporation, and vapor traps heat in the atmosphere, which adds to warming. But vapor also adds to cloud cover, which has two effects: it too traps heat at the earth's surface, but it also reflects heat from the sun back into space, which partially offsets warming. Scientists are not yet sure about the net effect, although their tentative judgment is that increased cloud cover probably raises temperatures. By contrast, there is little doubt that the accelerated melting of ice caps and glaciers tends to swell rivers and raise sea levels.

As experts combine what they know with what they suspect may happen, they can imagine Biblical-scale floods, droughts, and famines: Manhattan and much of Florida under water; breadbaskets turned into wastelands; a major change in the Gulf Stream that could bring Siberian winters to what is now temperate Europe; the buckling of Arctic permafrost that could release tens of billions of tons of methane, a greenhouse gas that is twenty times more powerful in its heat-trapping effect than carbon dioxide.[7]

These and other "perturbations"—the term scientists use for disturbances that result from changes in climate—would constitute not just an environmental and humanitarian disaster but a geopolitical one, particularly if they interact in ways that are mutually exacerbating. Defense and intelligence agencies of the U.S. government are concerned about global warming becoming a cause of political instability soon enough in the future to make it a factor in U.S. strategic planning today.[8]

BECAUSE THE PROBLEM OF CLIMATE CHANGE is almost certainly anthropogenic, the solution, insofar as there is one, must be the same: a change in human activity that counteracts—or, as we have learned to put the goal more modestly, mitigates—the consequences of our two-hundred-year accidental experiment.

Scientists and economists believe the world can stay below the 3.6°F/2°C threshold at a reasonable cost. That calculation takes into account the cost if we do *not* act and if the consequences of global warming are as severe as science indicates they might be. Precisely because there is uncertainty about where on the temperature scale the danger zone is, we should try to stay as much below the 3.6°F threshold as possible.

Moreover, we need to start reductions *now* in order to slow temperature rise later. Even if we could throw a switch and shut down all emissions, gases that are already in the atmosphere will continue to trap heat for some time to come. Once emitted into the atmosphere, a molecule of carbon dioxide, or CO_2, lingers for decades. So gases emitted today are added to ones that have been around for fifty years or more.

The current concentration of CO_2 in the atmosphere is about 385 parts per million (ppm), and growing by 2 ppm each year. The IPCC believes that if that level rises and stays above 400 ppm over the next several decades, the consequent increased warming could push global temperatures past 3.6°F by mid-century.[9]

So 400 ppm is—like 3.6°F—another dangerous ceiling, and it is one we are already close to hitting. Yet the production and consumption of energy from fossil fuels at current rates puts us on a course that may well boost CO_2 concentrations to nearly 1,000 ppm by 2050—more than double the level we must avoid.*

IF WE CONTINUE WITH BUSINESS AS USUAL, the globe could keep warming for millennia. Even if the human species is biologically resilient enough to survive for centuries, the human enterprise may well be hard to maintain in anything like its current form.

*This book refers primarily to CO_2 since it is the most prevalent anthropogenic greenhouse gas. There are five others. Nitrous oxide and methane are natural gases that, like CO_2, have been emitted into the atmosphere because of human activity. Sulfur hexafluoride, hydrofluorocarbons, and perfluorocarbons are synthetic, in that they do not occur in nature and are in the atmosphere because of industrial emissions. In addition, black carbon or soot (a particulate, not a gas) contributes significantly to the greenhouse effect.

Granted, that is the worst case. Granted, too, it is debatable. Scientists, like everyone else, can only guess about the future. The dangers the IPCC is warning of may be exaggerated, or—best case—they may not come to pass.

But we cannot count on the scientists overestimating the problem; it is just as likely they are underestimating it. In the past, the IPCC's reports were often criticized for equivocating or erring on the side of caution. Recently they have been attacked from the other direction after some reports were found to contain substantive errors and citations of work that had not been peer reviewed. Yet despite these controversies, most independent evaluations of the IPCC's two decades of work—including separate assessments by over fifty national science academies—have endorsed the panel's findings about the cause, nature, pace, and possible consequences of the climate change.

In any event, we do not want to sit back and wait for the future's verdict on whether the panel is alarmist or too sanguine. Nor can we wait for the science to be even more convincing before we take action. While the contributors to the IPCC and other researchers continue to refine their analysis and forecasts, governments should put in place policies that will keep CO_2 concentrations from reaching the barely safe limit of 400 ppm by 2050.

Today, humanity is cumulatively emitting, on a yearly basis, around 30 gigatons of CO_2. A gigaton is a billion tons. Thirty gigatons is about the weight of 8,000 Empire State Buildings, which, if stacked one on top of another, would reach almost 2,000 miles into space.

To keep CO_2 concentrations below 400 ppm and thereby keep temperature rise below 3.6°, we should use the next four decades to cut the current output of 30 gigatons a year approximately in half. So that is another target for mitigation: a staged process that would bring the global annual output down to 15 gigatons a year by 2050.

To reach that goal, we have to build a new worldwide system for generating and using energy. We have to begin quickly in order to achieve the bulk of the necessary cuts between 2020 and 2035 so that there is some hope that, by 2050, emissions will have come down to 15 gigatons, concentrations will have stabilized below the 400 ppm level, and temperature rise will have flattened out before hitting the 3.6°F mark.

At the heart of this mammoth undertaking is a transition from a high-carbon to a low-carbon global economy—that is, one that is powered as much as possible by forms of energy that do not burn fossil fuels and therefore do not pump CO_2 into the atmosphere.

THE TRANSACTION

The design of a mitigation strategy combines complicated science with elementary economics. The world relies on carbon-based fuels—principally coal, oil, and natural gas—because they are relatively cheap. But the price of fossil fuels does not begin to cover the costs incurred in damage to the environment. Hence, mitigating global warming requires a high price on carbon so that citizens, public utilities, companies, institutions of all kinds, and entire countries have an incentive to shift to alternative sources of energy on a vast scale and within a relatively short amount of time.

Richard Gephardt—an American politician who has grappled with the issue in government and the private sector—has said that changing how the planet generates and consumes energy will be "the single most difficult political transaction in the history of mankind."[10]

The United States must be part of that transaction. Of the dozen or so countries that have put most of the greenhouse gases into the air, the United States is far and away at the top of the list. With only about 5 percent of the world's population, it is responsible for about 17 percent of total accumulation of those gases in the atmosphere and for about 20 percent of the world's annual greenhouse gas emissions. (Of those 30 gigatons of CO_2 that will be emitted this year, just under 6 gigatons are from the United States.)

That dubious distinction—combined with the nation's unmatched power in the world economy and in the various institutions and arrangements that make up the international system—makes Americans more responsible than anyone else both for causing the problem and for leading the search for a solution.

Barack Obama takes that responsibility personally. He has often said that his presidency coincides with what may be a now-or-never moment because the peril is growing and the opportunity to avert it is shrinking.

Obama and members of his administration often speak of being at an inflection point.[11] That phrase is a mathematical term adapted to organizational theory. Business executives use it to describe how a new technology or a new source of competition forces a company or a sector to adjust and innovate. Andy Grove, the long-time CEO of Intel Corporation and a guru of corporate strategy, has defined an inflection point as an event that changes the way people think and act.[12]

We should be so lucky—and so smart—as to be at an inflection point. So far, we have begun to think more realistically, but we are still a long way from taking sufficient action.

For nearly twenty years, the nations of the world have been engaged, under the auspices of the United Nations, in a permanent floating negotiation on a global deal to reduce greenhouse gas emissions. That process nearly collapsed at the Copenhagen conference in December 2009. The chaos Obama found on arrival and the barebones agreement he was able to get before departing convinced him and many others that the next summit, scheduled for Cancun, Mexico, in late November and early December 2010, requires revised objectives and streamlined means of achieving them.

Foremost of those changes should be less reliance on the cumbersome UN-led pursuit of a legally binding global treaty, a process that has been on slow forward. Building on the work they have been doing together over the past year, the United States, the European Union, China, and India should form the core of an expanding circle of countries that develop their energy policies and regulate their emissions in an increasingly coordinated fashion.

As this book went to press in April 2010, the meteorologically memorable Washington winter had given way to the first signs of spring. Just weeks after the last traces of snow left over from the blizzards had melted, there were reminders that local weather can be deceptive about what is happening to global climate. Many species of birds were migrating north sooner than they had in past decades; the cherry trees along the Tidal Basin were blooming earlier than a few decades before; and data from around the world showed that, on a global basis, it had been one of the warmest winters on record.[13]

There was also a change in the air politically. In March, after a long, bitter, and debilitating battle over health care, Congress finally passed a bill that Obama signed into law—but with no Republican support in either the House of Representatives or the Senate.

That augured badly for bipartisanship of the sort that will be essential for legislative progress on climate change. It had taken the support of eight Republicans for Democrats to pass, by the narrowest of margins, a climate and energy bill in the House during the summer of 2009. That fall, counterpart legislation did not advance in the Senate. During the long, harsh, rancorous winter, the chances for climate legislation seemed bleak. Then, in the spring of 2010, the prospects improved somewhat when Senator Lindsey Graham, a Republican from South Carolina, began working with John Kerry, a Democrat from Massachusetts, and Joseph Lieberman, an Independent from Connecticut, in crafting a bill that would put a price on CO_2. That effort was, by all accounts, an uphill battle against both Republican and Democratic resistance. Still, it was the most serious effort yet in the Senate.

With summer approaching, it seemed that the U.S. domestic debate over the senators' initiative, followed by the Cancun conference in the late fall, might determine whether 2010 would be a true inflection point—a chance that we must not miss to avert a threat that is almost, but not quite, without precedent. As the issue is joined in the months ahead, it is worth recalling the only other time that humanity has confronted a comparable peril—and met the challenge of keeping it in check.

CHAPTER TWO

THE GOOD NEWS OF DAMNATION

IN THE SECOND HALF of the twentieth century, it took tens of trillions of dollars in military expenditures, a sustained combination of toughness and prudence as well as the indispensable element of luck to hold at bay the specter of a thermonuclear holocaust. The planet survived because the leaders responsible for the danger acted responsibly and effectively in preventing it. They understood that a scientific advance left them no choice but to think in a radically different way about national interests and the conduct of international relations.

The result was a new ethical and political logic appropriate to the Nuclear Age. With the advent of atomic power, the overarching objective of national security policy and diplomacy was to prevent a prodigy of human ingenuity from wreaking havoc on much of humanity and on the entire ecosphere. The threat was global, but the responsibility for averting it lay with a small group of big powers that engaged in the necessary combination of restraint and cooperation.

Much the same logic can now be adapted to the Age of Global Warming.

CIVILIZATION IS AN INTELLECTUAL work in progress, the result of the human capacity for reason, imagination, and innovation. That process has two dimensions. The first is technological: human beings' ability to understand natural phenomena, then to harness the forces of nature to practical purposes, such as using fire to cook and to survive the winter or inventing the wheel, the steam engine, or the nuclear reactor.

The second dimension is philosophical, ethical, or religious: human beings' need to understand their place and purpose in the cosmos, to define virtue, and to regulate individual and collective behavior. Politics and governance are an ongoing refinement of that code of conduct and of how it should be encouraged and enforced.

There has been a cause-and-effect dynamic, and therefore a lag, between these two drivers of civilization: technological advances have tended to precede and necessitate philosophical, ethical, and political advances. Primordial inventors probably figured out how to sharpen stones for the purpose of bashing heads before primordial ethicists made "thou shalt not kill" a rule of communal life. Down through the ages, human progress in controlling nature has been a mixed blessing and sometimes an outright curse, presenting new dangers that required new ways of controlling human nature.

Ethical progress has been slow in coming, often to the despair of great thinkers. Contemplating yet another outbreak of mayhem in Europe in the late eighteenth century, Immanuel Kant chafed at the seeming uselessness of philosophy—his life's work—in the face of the advances in chemistry, physics, and metallurgy that made warfare all the more brutal. "It seems ridiculous," Kant lamented, "that while every science moves forward ceaselessly, metaphysics, claiming to be wisdom itself, whose oracular pronouncements everyone consults, is continually revolving in one spot, without advancing a step."[14]

In the early nineteenth century, Friedrich Hegel found politically motivated violence hard to reconcile with his progressive theory of history. The violence and destruction of war, he wrote, were "not the work of mere Nature, but of the Human Will—a moral embitterment—a revolt of the Good Spirit (if it has a place within us)."[15]

A century and several wars later, Sigmund Freud wrote in *Civilization and Its Discontents,* "The fateful question for the human species seems to me to be whether and to what extent its cultural development will succeed in mastering the disturbance of its communal life by the human instinct of aggression and self-destruction. . . . Men have gained control over the forces of nature to such an extent that with their help they would have no difficulty in exterminating one another to the last man."[16]

UNTIL THE MIDDLE OF THE TWENTIETH CENTURY, history followed a predictable and doleful script: new, ever-more deadly techniques for waging war were put to use with effects so horrible that they prompted new, improved, but still highly imperfect—and very different—ways of keeping the peace. Some were formal legal pacts, others were more informal arrangements among the great powers, and the most effective were a mixture of the two.

In the seventeenth century, the carnage of the Thirty Years' War, which reduced the population of central Europe by a third, induced the combatants to enter into the Treaty of Westphalia. Instead of big powers cutting a deal that gave each a sphere of influence over weaker neighbors, the Westphalian system established a modus vivendi among a multitude of sovereign states and principalities, large and small, within a system based on the twin principles of self-governance and mutual interest.

A century and a half later, the Napoleonic wars ended with the Congress of Vienna and the Concert of Europe, which established a new system based on the balance of power among the major states of the time. That system broke down with World War I, which ended at Versailles, where the victors tried to preserve the formal sovereignty of all nations within a system of collective security that the League of Nations was supposed to enforce. The abysmal failure of that effort contributed to the onset of World War II.

After that conflagration, the allies and others agreed, at an international conference in San Francisco in 1945, to create a system that combined features of Westphalia (the primacy of the nation-state), Vienna (balance of power), and Versailles (collective security). The United Nations General Assembly represented the formal acknowledgment of all member states' "sovereign equality," while the Security Council concentrated special authority in the hands of the United States, two key western European countries (Britain and France), the Soviet Union, and China.

That hybrid might have proved no more durable than the earlier models from which it drew were it not for an entirely new factor: nuclear weaponry. Reactive peace gave way to proactive peace—a peace that did not just clean up the mess of the last war but prevented the next one. It was not the remembrance of World War II that stimulated that breakthrough—it was the prospect of World War III.

Freud's fear had come true: humanity now had the capacity to obliterate itself. The United States and the Soviet Union were the two most powerful states of all time because of their military capability. They were also two of the most antagonistic states. At any earlier point in history, they might have gone to war, perhaps repeatedly. Yet because of the form war between them would take, it was not an option. For them, the most basic of human instincts—survival—required that Clausewitz's famous maxim be turned on its head: peace would now have to be the conduct of war by other means.

That peace was called, with deliberate irony, the cold war. The United States and the Soviet Union competed by means that were sometimes deadly (spies versus spies, proxies versus proxies), but their governments cooperated in avoiding direct military conflict.

IN 1977, THIRTY YEARS into what John F. Kennedy called the "long twilight struggle" of the U.S.-Soviet showdown, the political philosopher Michael Walzer wrote, "Nuclear war is and will remain morally unacceptable, and there is no case for its rehabilitation. . . . Nuclear weapons explode the theory of just war. They are the first of mankind's technological innovations that are simply not encompassable within the familiar moral world."[17]

Five years later, in 1982, Jonathan Schell, a staff writer for *The New Yorker,* took Walzer's proposition to a higher level of passion and impact with a series of three articles that were then assembled into a book, *The Fate of the Earth.* As the title suggests, the work was a combination of sermon and manifesto. A full-scale thermonuclear "exchange" between the United States and the Soviet Union would devastate humanity through the combination of the war itself, the aftereffects of radiation, and the contamination of the planet, leaving in its wake, in Schell's phrase, "a republic of insects and grass."[18]

Schell called not just for the repudiation of the doctrine of mutual deterrence and the abolition of nuclear weaponry but for the rejection of national sovereignty and the abolition of nation-state. "The task," Schell concluded, "is nothing less than to reinvent politics: to reinvent the world." The book became an instant international best seller.[19]

The following year, the U.S. Conference of Catholic Bishops released a manifesto of its own. Known as a pastoral letter, it was titled "The Challenge of Peace—God's Promise and Our Response." It too received an unusual degree of attention, resonating with people who otherwise had little or no connection with the Catholic Church and its admonitions—and who, in many cases, had not thought much about nuclear deterrence or the war it was meant to deter.[20]

The "letter" (which, at 43,000 words, was considerably longer than this book) was an indictment of the arms race as "a folly which does not provide the security it promises" and a call for a "moral about-face" in the form of "immediate, bilateral, verifiable agreements to halt the testing, production, and deployment of nuclear weapons systems." The leaders of the world must "summon the moral courage and technical means to say no to nuclear conflict; no to weapons of mass destruction; no to an arms race which robs the poor and the vulnerable; and no to the moral danger of a nuclear age which places before humankind indefensible choices of constant terror or surrender."

Schell and the bishops acknowledged that the U.S. and Soviet leaders of the time were not entirely oblivious of the danger. As Schell put it, "I believe that without indulging in wishful thinking we can grant that the present leaders of both the Soviet Union and the United States are considerably deterred from launching a nuclear holocaust by sheer aversion to the unspeakable act itself."

In fact, the moral and political implications of nuclear weapons had been intuitively evident to a number of leaders responsible for their development and deployment even before a prototype of the atomic bomb was exploded in the sands of New Mexico in 1945. On the eve of the first American test, Henry

Stimson, the secretary of war, confided to his diary, "The world in its present state of moral advancement compared with its technical development would be eventually at the mercy of such a weapon. In other words, modern civilization might be completely destroyed."[21]

Stimson understood, as did Hegel and Kant, that the time would come when practitioners of politics would have to revise and reinforce the ethical foundations of their profession. Otherwise advances in the technology of destruction would threaten the whole edifice of civilization. Stimson realized that time had come. So did his boss. Shortly after the test, Harry Truman wrote in his own diary, "We have discovered the most terrible bomb in the history of the world. It may be the fire destruction prophesied in the Euphrates Valley Era, after Noah and his fabulous Ark."[22]

In 1947 the editors of the *Bulletin of the Atomic Scientists,* which had been founded two years before by scientists, engineers, and others who had worked on the Manhattan Project, tried to focus public attention on the danger of nuclear war by adding a regular feature: a Doomsday Clock that conveyed the editors' concern about how close the world was to disaster. And that was when only the United States had the A-bomb.

Two years later, when the Soviet Union ended the American monopoly, its leaders were under no illusion about what war would mean. In the wake of the first Soviet test, Joseph Stalin remarked to an aide, "Atomic weapons can hardly be used without spelling the end of the world." In 1953 one of Stalin's successors, Georgi Malenkov, said publicly that a war waged with "modern weapons" would be "the end of world civilization."[23]

A prominent outlier was Mao Zedong, the founding leader of the People's Republic of China. In a 1957 speech in Moscow Mao boasted that "even if half of the population in the world died [in a nuclear war], the other half would survive. Moreover, imperialism would be destroyed and the entire world would be socialized."[24]

Mao was flattering his Soviet hosts in a way that probably made them wince—and that made the world glad he did not have a bomb of his own. Even when China joined the nuclear club in 1964, most of the world's weapons were in the hands of the United States and the USSR. By then the successors of Truman, Stimson, Stalin, and Malenkov were already thinking about how to do what Schell would call for in *The Fate of the Earth:* the reinvention of international politics—though not in the way Schell advocated. To ensure that their governments did not stumble into war by mistake, they worked out a system for regulating the arms race that would stabilize deterrence.

For decades delegations of U.S. and Soviet officials and military officers sat across from each other and haggled over the composition and capabilities of each side's arsenals. They first limited and then reduced the numbers, types, and modes of deployment of nuclear weapons in ways that diminished the danger of a war erupting as a result of surprise attack, accidental launch, or miscalculation.

In both governments, arms control became a prestigious career track. The institutionalization of nuclear diplomacy turned out to be durable enough to withstand the intrinsic animosity of the U.S.-Soviet relationship and the vacillations of American policy as presidents' popularity rose and fell and administrations came and went.

What started as a joint venture between superpowers went global when much of the rest of the world signed on to the Non-Proliferation Treaty (NPT), which entered into force in 1970 and helped prevent the spread of nuclear weapons to countries that did not already possess them.

In this respect, the cold war provided a precedent for how the world can most expeditiously deal with climate change: start with the countries most responsible for the problem; then, once they have taken responsibility for the solution, bring other countries into a web of agreements and coordinated initiatives.

THE BASIC SIMILARITY between the Nuclear Age and the Age of Global Warming is that, in both cases, a potentially existential threat has imposed an existential imperative—a priority that trumps all others. From the late 1940s until the late 1980s, Americans did not *have* to push for the creation of the United Nations, rescue Europe, put a man on the moon, or even contain communism. But Americans did have to manage their relationship with the Soviet Union in a way that prevented nuclear war. In fact, the United States was able to integrate all those other ventures—creating strong international institutions, promoting security and prosperity in Europe, establishing technological supremacy, and competing with Soviet power and influence—into a strategy that had, at its core, the primary goal of preserving the nuclear peace.

That was relatively easy to do because during the cold war people had a clear idea, based on past experience, of what a hot war would be like. The United States had waged nuclear war against the Empire of Japan for four days in August 1945. The Sunday after the incineration of Hiroshima, Robert Maynard Hutchins, then the chancellor of the University of Chicago, spoke of "the good news of damnation": the headlines, he hoped, would frighten the leaders of the world into taking the steps necessary to avert catastrophe.[25]

Those of us in our sixties can remember, from childhood, photographs in *Life* magazine and elsewhere of Japanese with disfigured faces, burned bodies, and in many cases, only days to live before they would die of radiation poisoning. In the years that followed, we watched film clips of mushroom clouds billowing skyward in the American desert and H-bombs obliterating atolls in the South Pacific. The regular duck-and-cover drills in which we took part in elementary school, on the farfetched premise that such precautions would protect us from a nuclear blast, were an education unto itself for us and our parents. What we remembered from the 1950s conditioned us to take the appeals of Jonathan Schell and the Catholic bishops all the more seriously in the 1980s.

In marked contrast to climate change, there was comparatively little debate about the science, facts, or consequences of nuclear war.

THERE ARE TWO OTHER salient differences between our survival of the cold war and what we must do in response to global warming.

The first is essentially Hutchins's point about the saving grace of looking damnation in the face. Nuclear war concentrated the mind even more effectively than the prospect of being hanged in a fortnight. Never mind a fortnight: nuclear war was a permanent possibility of something that could happen at any time. That threat was—and, still is—a Sword of Damocles.

By contrast, the potentially lethal effect of climate change is more like the Chinese method of execution known as slow-slicing, or, literally, death by a thousand cuts. That difference gives us time that we did not have during the Cuban missile crisis, but it deprives us of a sense of urgency, a sense that our lives are in immediate danger.

The second difference is the degree of control we have over what happens. With climate change, we have *some* control; we must hope—and act on the hope—that we can do enough to protect our progeny from the worst. But we do not have anywhere near as much control as we had—and still have—in preventing nuclear war. Every day for four decades leaders in the White House and Kremlin had the option of pressing a button that would send the missiles flying to blow up much of the world. But, also every day, they had the option of *not* pressing it. So preserving the nuclear peace was a matter of our leaders' giving us a stay of execution a day at a time, waking up every morning and simply not doing something colossally stupid.

That is still a fact of our lives, and it will remain one for a very, very long time—who knows: maybe even long enough for future leaders to abolish nuclear weapons and reinvent the world in the way Schell would have had us do thirty years ago.

With climate change, however, we have already done something stupid. Without meaning to, we, the human race, pressed the button that triggered the problem two hundred years ago. Every day that we do not do something exceedingly smart is a day we spend with the slow-slicer.

CHAPTER THREE

THREE PRESIDENTS AND A PROCESS

SO WHAT IS THE SMARTEST thing we can do to mitigate climate change? The answer is easy: slow, stop, and then reverse the rate at which we—all 6.8 billion of us—are emitting greenhouse gases. What sort of deal will it take to accomplish that goal, and who should broker it? For twenty years, the answer has been a legally binding treaty that is to be negotiated through a process, run by the United Nations, that will commit all the major emitters to a strict schedule of deep reductions. The trouble is, that process is not working. The greater the effort, the more elusive the goal; and the more powerful the nations involved, the more they have resisted a treaty that imposes legally binding restrictions on them.

The United States is a case in point. Some of the best climate scientists in the world are Americans. So are some of the most eloquent and influential advocates for a global compact. Moreover, global compacts are an American specialty. At pivotal moments in the twentieth century, U.S. presidents were the chief architects, master-builders, and principal funders of international institutions—notably, the UN itself—that constitute a rule-based world order and have promoted the prosperity and security now threatened by climate change.

Yet in the face of that threat, all four of America's most recent presidents—including, up until now, its current one—have, for one reason or another, been unable to rise fully to the challenge. Part of the reason is domestic politics. The nation's chief executive shares responsibility for success and failure with his partners in government at the other end of Pennsylvania Avenue, on Capitol Hill. Beyond the frictions between the two major political parties and the two branches of government, there has been another, more basic impediment: the difficulty of reconciling the diplomatic ideal of a globally binding treaty with the political reality of national sovereignty and economic self-interest.

That is not by any means a uniquely American dilemma—it is a universal one. All 192 nations on earth are grappling with some version of it. A number of them have emerged as major international players in the two decades since climate change got the world's attention, and their participation in a global deal, whatever form it takes, is essential to its efficacy.

One reason that there was so little progress during the twenty years that George H. W. Bush, Bill Clinton, and George W. Bush were in office was that the diplomacy of climate change was a UN-driven—and UN-limited—process. Whether Barack Obama remains similarly constrained by that process is one of the big questions hanging over his presidency.

AN ENVIRONMENTAL PRESIDENT

The senior Bush was the first president to hold office after the cold war ended and the first to recognize that the Age of Global Warming had begun. That timing positioned him for leadership on climate change. So did his background and attributes as a statesman. As Richard Nixon's ambassador to the United Nations in the early 1970s, Bush found diplomacy better suited his temperament and operating style than the rough-and-tumble of electoral politics. He quickly established a reputation for being unusually collegial and respectful of other governments' perspectives. Almost every day he would make a point of calling on two or three foreign counterparts to gauge what was happening in their part of their world and ask them for their reaction to U.S. policy.

Bush's next assignment—as President Gerald Ford's envoy to Beijing—gave him a chance to get to know a country that would soon emerge as a major power on many issues including climate change.

When Bush ran for the presidency in 1988, climate was in the news. It was, up to that point, the hottest year on record. Bush came into office promising to be an "environmental president" who would counteract "the greenhouse effect" with "the White House effect."

Once in office, Bush signed into law the U.S. Global Change Research Act, which authorized and funded valuable scientific work, and the Clean Air Act of 1990, which was intended to reduce air pollution and acid rain caused by sulfur dioxide and other particulates. While the Clean Air Act did not deal directly with climate change, it included two provisions that might some day be used to do so.

First, even though carbon dioxide was not explicitly identified as a potential danger, the new law gave the Environmental Protection Agency (EPA) broad authority to regulate any pollutant that "may reasonably be anticipated to endanger public health or welfare." Second, the Clean Air Act regulated the output of sulfur dioxide from coal-burning power plants in an innovative way, called emissions trading, that might some day be applied to CO_2 and other greenhouse gases. The law empowered the EPA to issue permits to companies, allowing them to emit set amounts of sulfur dioxide. Over time the level would have to come down. Companies that could not reduce their emissions fast enough could buy permits from companies that were able to reduce quickly.

This cap-and-trade feature was denounced by some environmentalists as granting "pollution rights." Some industry leaders feared that the system would be too complicated and expensive. Both fears proved wrong. The new law succeeded in cutting emissions that caused acid rain with far less disruption and expense than had been predicted even by its proponents.[26]

A number of Bush's colleagues believed he might have classified CO_2 as a pollutant that should be capped if he had followed his personal instincts and the advice of William Reilly, the head of the EPA. Bush and Reilly constituted what was sometimes called the "green faction" in debates at the White House.

John Sununu, Bush's chief of staff, was the most powerful voice on the other side. A former engineering professor, Sununu was aware of reports projecting dire consequences if current trends continued. He was extremely skeptical about the science behind these warnings. He tried to ban the phrases "climate change" and "global warming" from administration statements, a directive Reilly ignored. Sununu was willing to support the UN's establishment of what became the Intergovernmental Panel on Climate Change—but for reasons that he felt would serve his purpose, not Reilly's. Given how hard it was for the scientific community in any country to come to agreement on anything, Sununu was confident that an international consensus would be impossible on an issue as complex as climate change, and therefore the pressure would be off the United States do anything.

There were also political reasons to oppose limits on CO_2. The Republican base, big business, and social conservatives were against more government regulation of emissions.[27]

IN JUNE, 1992—five months before the election that Bush hoped would give him a second term—the United Nations convened a two-week "Earth Summit" in Rio de Janeiro. With 108 heads of state and government in attendance, Rio was, at the time, the largest such gathering in history—vivid evidence that climate change was now receiving global attention at the highest level.

Many of the leaders present—particularly the Europeans—had hoped that the outcome would be a treaty establishing legally binding, country-specific targets and deadlines for reducing greenhouse gas emissions. But the U.S. delegation, headed by Bush, balked at binding targets. Controversial as the Clean Air Act had been in the United States, it was national legislation. That made it hard to portray as an infringement of American sovereignty, whereas a Rio treaty with legally binding cuts would, if ratified, subject U.S. domestic energy use to international regulation under UN auspices.

Largely because of U.S. opposition, the final treaty—known formally as the UN Framework Convention on Climate Change—was modest and vague. In signing it, Bush committed the United States to little more than participating in subsequent conferences to review progress and assess whether binding targets could be justified.

The results in Rio disappointed and angered environmentalists around the world, and the United States came in for much of the blame. Bush was harshly criticized by Senator Al Gore, who had just published a book on environmental degradation with special emphasis on climate change. *Earth in the Balance: Ecology and the Human Spirit* quickly made the best-seller lists and energized public opinion much as Jonathan Schell's *Fate of the Earth* had done during the cold war.

Gore's book made a strong impression on Bill Clinton, who later said it was one of the main reasons he brought Gore onto the Democratic ticket with him. Once they were in the White House together, environmental issues and particularly global warming were often the subject of their weekly private lunches.[28]

The new administration tried to link the reduction in greenhouse gases to a reduction of the federal budget deficit by increasing the tax on gasoline in order to raise revenues for the government while encouraging consumers to drive more fuel-efficient vehicles. The budget bill that Clinton signed into law in 1993 added 4.3 cents per gallon to the federal gasoline tax and gave congressional Republicans ammunition to attack him as a tax-and-spend Democrat. Along with the failure of the administration's health care initiative, the gas tax contributed to the Democrats' loss of control of both houses of Congress in the 1994 mid-term elections.

A TROUBLESOME MANDATE

In March 1995 the signatories to the Rio Framework Convention met in Berlin to give more clarity of purpose to the next stage of negotiations. Clinton was more willing than Bush had been to consider binding emissions cuts, but having paid a political price the previous year for his gas tax, he did not want to get into a negotiation over specific targets until he got past his own reelection campaign in 1996. Therefore the negotiation was postponed until a climate summit scheduled for Kyoto in 1997.

The U.S. delegation in Berlin did, however, agree to another proposal that would, in the years to come, bedevil both the diplomacy and the U.S. domestic politics of climate change: countries designated *developed* would start negotiating binding obligations that would not apply to those designated *developing*.

The distinction between the two categories, along with the principle of "common but differentiated responsibilities," had been part of the Rio treaty.* The Berlin meeting turned that principle of special treatment for developing countries into an outright exemption from binding targets. The argument for doing so was rooted in an historically based and economically motivated resentment on the part of countries that had emerged from colonialism and had yet to recover from decades, if not centuries, of exploitation. Developed countries bore historical responsibility for climate change, and much of the world's wealth was still concentrated in those countries, while most of the world's poor were still in the developing world. Now that developing countries were finally catching up economically with the established industrial powers, they were determined not to be losers on the playing field of globalization.

That case was more compelling in 1995 than it is today. Recent years have shown that the winner-loser, rich-poor divide does not exist solely between blocs of countries but *within* countries as well. Many Americans feel like losers, especially if they have lost manufacturing or service sector jobs to factories or call centers in China or India. And a growing number of Chinese and Indians

*The UN Framework Convention on Climate Change created at the Rio summit in 1992 actually divided countries into three categories: 1) developed countries in Annex I (those with highly industrialized economies or with economies that were in transition from communism); 2) developed countries in Annex II (those that would help pay the costs of developing countries in adapting to climate change—which were many of the same countries in Annex I, minus those in transition from communism); and 3) developing countries.

feel increasingly like winners. Their countries are not just "developed" in significant ways—they are highly competitive in many fields and well on their way to being dominant in several.

Like climate change itself, the economic and political dynamism of what had been the poorer countries of the world—indeed, their *development*—was on fast forward to an extent that few of their own leaders fully understood. Part of the price for their success in closing the prosperity gap was that they were closing the emissions gap as well. The faster they caught up with the developed countries in being part of the problem of climate change rather than just victims of it, the less differentiation they could reasonably expect in their responsibility for a global solution. But once the principle of "differentiated" legal obligations for the two categories of countries was enshrined in what became known as the Berlin Mandate, the developing countries would cling to it for years to come.

In June 1997 President Clinton presided in Denver over the annual meeting of the Group of Eight, or G-8—the heads of state and government of Canada, France, Germany, Italy, Japan, Russia, the United Kingdom, and the United States. Now that Clinton had been reelected, the Europeans tried, once again, to get him to join them in pushing for a treaty that would feature binding emission commitments as the goal for Kyoto later that year.

Clinton, who still had a Republican-controlled Congress to contend with, refused. The Denver communiqué promised only that the Kyoto summit would produce "meaningful, realistic and equitable targets that will result in reductions of greenhouse gas emissions by 2010." The Europeans left Denver visibly and vocally miffed. They did not like the trio of adjectives: "meaningful" sounded ambitious, but "realistic" sounded less so, and "equitable" could be used either to support or dilute the Berlin Mandate. They feared that the results in Kyoto, like those in Rio, would be underwhelming—and that American foot-dragging would, as in Rio, be one of the principal reasons.

Clinton later told his staff that he felt he had been "the skunk at my own garden party." He saw himself as a conciliator and hated being cast as an obstructionist—especially on the issue of climate change. Yet because of congressional adamancy, the United States seemed destined to be the skunk in Kyoto as well.

A MONTH AFTER DENVER, a bipartisan group of senators made sure that would happen. A resolution cosponsored by Robert Byrd, a Democrat from West Virginia, and Chuck Hagel, a Republican from Nebraska, put U.S. negotiators on notice that the Senate had two criteria for deciding whether to ratify whatever treaty emerged from Kyoto. First, nothing in it could be construed as threatening "serious harm to the United States economy, including significant job loss, trade disadvantages, increased energy and consumer costs, or any combination thereof." Second, the Kyoto accord must back away from the Berlin Mandate's promise to differentiate between obligations for developed and developing countries.

The Byrd-Hagel resolution passed 95-0. For years afterward, Clinton would often say, "Kyoto was the only bill I lost before I sent it to the Congress."

MANY EUROPEANS REGARDED CLINTON'S domestic opposition with a touch of schadenfreude. For decades, the United States had been in the lead on most major international—and especially Euro-Atlantic—diplomatic ventures. Climate change gave the Europeans an opportunity to step into that role themselves.

Their ambitions for Kyoto were also a reflection of their success in economic and political integration after World War II. That process—or as the Europeans called it, their "project"—began with France and Germany resolving to bind their economies so tightly that they would never again go to war. They did so through the six-member European Coal and Steel Community, which, by 1992, had evolved into the European Union, with fifteen member states. In Kyoto those nations constituted a solid bloc that was far ahead of the rest of the world on climate change mitigation. In the months leading up to the conference, the EU environment ministers had adopted a continentwide goal to cut greenhouse gas emissions 15 percent below 1990 levels by 2012. They hoped to convince the United States to adopt a similar target.

The designation of 1990 as a benchmark was, in one respect, reasonable, since it was around that year that climate change became widely recognized as a global threat. But the Europeans had another, less objective reason for picking 1990: it gave them an advantage vis-à-vis the United States. At the end of the previous decade, France had ramped up its reliance on nuclear reactors instead of coal-fired plants; in the United Kingdom, Margaret Thatcher, in her determination to break the political power of unions, had shut down coal mines; and in Germany, the collapse of the Iron Curtain and the Berlin Wall allowed the Federal Republic to phase out the inefficient, dirty, coal-based industries of the defunct East German state. As a result, in the months leading up to Kyoto in 1997, Europe's emissions were only 1 percent above where they had been seven years before, making 1990 the ideal baseline for the European Union to use in setting its targets for 2012.

By comparison, the U.S. economy had expanded dramatically throughout the 1990s, and with that boom had come a sharp rise in energy use and an 11 percent increase in greenhouse gas emissions.

Europe had another advantage as well. Its system of parliamentary democracy and proportional representation gave environmental activists political power well beyond their numerical strength. In several key northern European nations—Germany, the Netherlands, and the Scandinavian countries—electoral rules made it possible for minority parties with strong environmental platforms but with only about 10 percent of the popular vote to join the government with as much as 20 percent of the coalition's parliamentary votes. With that clout came cabinet positions, often including the environment ministry, which further helped them make a national priority of their parties' defining issue.

Green parties around the continent coordinated among themselves and with allies from other parties, further leveraging their influence over EU policy. Because protection of the environment was broadly popular, regardless of the composition of governments, those that had pledged aggressive climate targets were able to shame less enthusiastic neighbors into taking action.

An overarching factor conducive to a coherent, forward-leaning European policy in Kyoto was a profound political transformation in a region that had been ravaged by a major war every generation since the seventeenth century. In the aftermath of World War II, the birthplace of the sovereign nation-state had embraced the idea of "pooled sovereignty."

That phrase could barely be uttered in respectable political company in Washington. The American people and their elected representatives have, by and large, been highly protective of the sovereignty proclaimed in the Declaration of Independence and achieved in the Revolutionary War. To ensure that the United States would not enter into binding agreements with other countries lightly, the Founders wrote into the Constitution a provision requiring that treaties be ratified by a two-thirds majority of the Senate.

Hence a major difference between Europe and the United States: while European parliamentary and coalition politics empowered green minorities to force through bold climate policies, the rules of the U.S. Senate empowered minorities to block action. A fairly recent practice, without any basis in the Constitution, required sixty votes out of a hundred to prevent a filibuster on virtually any piece of legislation. The so-called supermajority was a misnomer, since it actually created a super *minority* when it came to blocking action. As a result, a president trying to pass legislation to curb greenhouse gas emissions would need to run a Senate gauntlet that was rigged to favor his opponents.

As the Kyoto summit approached, Clinton unveiled the U.S. negotiating position in a speech at the National Geographic Society. The United States would commit itself to binding emissions cuts that returned the nation to 1990 levels by 2012, with further cuts after that. This would have amounted to a 10 percent cut below where emissions were in 1997.

The pledge was immediately criticized by all sides. Europeans, who were promising 15 percent cuts below 1990 levels, seized on what they saw as a fresh reason for feeling superior, while American environmental groups saw their government's offer as vague, weak, and discouraging. "The air went out of the balloon," said John Adams, head of the Natural Resources Defense Council. "The policy announced today will not go nearly far enough to get us out of harm's way."[29]

Opposition from the right was, if anything, more scathing. "The Clinton-Gore global climate plan threatens the quality of life for all Americans," said Jim Sensenbrenner, an influential Republican U.S. representative from Wisconsin. The restrictions Clinton was proposing would "result in higher energy prices for working Americans and send U.S. jobs overseas."

A coalition representing oil companies, electric utilities, automobile manufacturers, and farm groups launched a multimillion dollar advertising campaign against the prospective treaty.[30] Meanwhile, several senators, including Chuck Hagel himself, prepared to attend Kyoto to make sure that the U.S. negotiators remembered the provision of the Byrd-Hagel resolution that was aimed at the Berlin Mandate, especially now that China was well on its way to displacing the United States as the world's No. 1 emitter (that happened in 2008), and India was heading toward the No. 4 position behind the United States and the European Union.

In an effort to break the impasse over differentiation, the United States was willing to establish a new system that would foster investments in clean energy projects and the protection of forests in developing countries. The Clinton administration calculated that perhaps, in return, key developing countries would agree to set goals of their own.[31] That hope was promptly dashed. The developing countries were glad to accept the help but not the condition. National targets, they feared, would limit their growth and infringe on their sovereignty.

The U.S. executive branch and the governments of other developed countries faced a choice in Kyoto between a flawed agreement and none at all. Rather than let the process collapse, they agreed to binding cuts by 2008–12, with the European Union pledging to reduce emissions to 8 percent below 1990 levels, and the United States pledging to reduce to 7 percent.

Clinton signed the Kyoto Protocol, calling it a "good first step." But in the face of certain rejection by the Senate, he announced he would not submit it for ratification until there were supplementary steps that entailed "meaningful participation of key developing countries"—a condition that would not be met in the years that followed.

THE BUSH BOYCOTT

George W. Bush ran for president in 2000 on a platform that suggested he, like his father, would be an environmental president. His policy prescriptions included treating CO_2 as a pollutant and instituting a cap-and-trade system for greenhouse gas emissions. Once elected, he appointed former New Jersey governor Christine Todd Whitman as administrator of the Environmental Protection Agency and Paul O'Neill, the chief executive of Alcoa, as treasury secretary. Both saw climate change as a real and growing threat and advocated a strong U.S. response. Whitman moved quickly to assure her counterparts in Europe that the United States would take steps to regulate CO_2.

Then Bush did an about-face. Less than two months into his administration, he backed off his campaign pledge on CO_2. Bush put Vice President Dick Cheney in charge of a task force to explore the full range of issues and options related to energy policy. Whitman and O'Neill were kept at arm's length.[32] Much like John Sununu twelve years before, White House political appointees tried to make the phrases "climate change" and "global warming" taboo: in 2001 they were scorned as "Gore talk."

In March 2001 the White House spokesman, Ari Fleischer, announced that an international effort like the Kyoto process was "not in the United States' economic best interest." Three months later, in a Rose Garden ceremony just before leaving on a trip to

Europe, Bush said that Cheney's review had led him to conclude that "the Kyoto Protocol is fatally flawed in fundamental ways."

The statement infuriated European governments and public opinion. At several stops on his trip, Bush encountered protestors brandishing posters calling him the Toxic Texan and depicting him with devil's horns. His official hosts for the most part swallowed their irritation, but they let him know that their own commitment to Kyoto was very much alive.

WHILE THE EUROPEAN UNION as a whole was determined to reduce its emissions by 2012 to a level 8 percent below where they had been in 1990, some nations—notably France, Germany, and the United Kingdom—set for themselves more ambitious national targets. Each pledged cuts of 20 percent. Others, particularly in the south—Greece, Portugal, and Spain—gave themselves a break so that they might close the economic gap with their northern neighbors.[33]

This framework had been settled long in advance of Kyoto. In some ways the Europeans had adopted, on a regional basis, the concept, as established in Rio, of differentiation between developed and developing countries, but with the important difference that no EU member state had a Berlin Mandate–like exemption from binding targets. Instead, poorer countries were allowed to have a lenient "growth target"—that is, one that let them reduce their emissions less rapidly than their more prosperous and advanced neighbors.

BACK IN WASHINGTON, BUSH AND CHENEY'S hard line on climate did not entirely stifle government research into the subject. Scientists and other experts at the EPA, the National Oceanic and Atmospheric Administration, the National Aeronautics and Space Administration, and the National Science Foundation continued to study the problem assiduously and issue reports raising concerns. However, an annual report from the executive branch to Congress was subjected to heavy editing by political appointees and, on some points, outright censorship.[34]

Meanwhile, the intelligence and defense communities were increasingly focused on climate change as a serious threat to the nation's security and the prospects for international peace. It was not an EPA report but a Pentagon study, commissioned in 2003, that warned of a "significant drop in the human carrying capacity of the earth's environment." Strategists incorporated into their "risk scenarios" the possibility that the melting of the North Polar ice cap would create conditions for a politically troublesome, potentially dangerous scramble for natural resources in the Arctic. A resurgent Russia might be tempted to throw its weight around at the top of the world. Future flooding or parching of arable land could cause vast waves of "climate refugees" and turn strong states into weak ones and weak states into failed ones. Failed states are often outlaw states, sources of regional instability, incubators of terrorism, and bazaars for lethal technology. Whole nations could be thrown into economic and political chaos, with all that might portend for internal and cross-border violence.

WHILE WHITE HOUSE OFFICIALS were trying to avoid discussing climate change, much of the country was stepping up action on the issue.

More and more families were opting for energy-efficient cars, switching to green technology for appliances and heating systems, and paying premiums for low-carbon products.[35] By the end of Bush's second term, nearly forty U.S. states had adopted emission-control measures. Scores of cities, from Seattle to Boston, sought to lower their carbon footprints by promoting energy efficiency in buildings, adopting low-carbon strategies for mass transport such as light rail, and experimenting with "smart metrics" for gauging—and better conserving—the usage of energy, water, and gas.

This environmentally conscientious policymaking at the local and regional level has been a useful first step in cutting emissions, and it has raised public awareness that saving energy is both important and relatively easy. But without an overall federal framework, the result is a crazy quilt of carbon-pricing mechanisms, emission regulations, subsidies for renewables, and other well-intentioned measures that will become increasingly uncoordinated, cost-ineffective, and in some cases mutually undermining.

Economists tend to concur that the federal government must take the lead for climate policy to be effective and must do so in a way that cuts across all relevant sectors of the economy and society.[36] Corporate leaders largely agree. The last thing an American company wants is to be forced to modify a product in fifty different ways to comply with fifty different state-level regulatory standards. While some corporations would prefer no regulation at all, many argue for uniform regulations that will create economies of scale in production and allow them to sell a product that meets uniform national standards in the largest possible market.

A moment of exquisite irony occurred in 2006, when the state of Massachusetts—then under the governorship of Mitt Romney, a Republican—petitioned the EPA to allow it to regulate auto emissions. The agency refused, arguing that piecemeal state legislation was not the most effective way to fight climate change and that only a legislated nationwide remedy would work, preferably in the context of a global regime—two conditions that Bush was, to say the least, doing nothing to bring about.

IN HIS SECOND TERM, Bush had somewhat moderated his position, although his policies were still a long way from those he had espoused during his campaign in 2000. He acknowledged what most of the rest of the world—again, the Europeans in particular—had long accepted: "The surface of the earth is getting warmer and the increase in greenhouse gases is contributing to the problem."[37] Part of the reason Bush moved closer to the view of the Europeans was his need for their support in dealing with Iraq, Iran, and Afghanistan. Still, he was careful not to associate himself with alarm about the severity and urgency of the phenomenon or calls for concrete action.

Then, in 2007, the IPCC—the international body that the president's father and John Sununu had helped bring into being—released its fourth comprehensive assessment. It reflected more confidence than earlier reports about the warming of the planet. It asserted that there was only a one-in-ten chance that warming was produced solely by natural causes and that most of the increases during the twentieth century were very likely the result of human-generated greenhouse gases.

The difference between the IPCC's conclusions and what Bush had finally acknowledged was significant. Bush admitted that human beings were contributing to the problem, but he left open to debate how *much* they were contributing and therefore how much human remedial action was required. The IPCC, by contrast, assigned humanity *most* of the blame and, by extension, considerable responsibility for cutting emissions to stem future warming. That responsibility would only grow—and become more difficult to discharge—given the projected increase in emissions over the course of the current century.

THE IPCC REPORT MADE an impression around the world. The editors of the *Bulletin of the Atomic Scientists* were still running a Doomsday Clock on the cover. To take account of global warming as well as nuclear proliferation (especially in Iran and North Korea), they set the minute hand at five minutes before midnight—two minutes closer than it had been during the Cuban Missile Crisis of 1962.

Other countries took notice as well, including in the developing world. None was more important than China, where the effects of climate change were already evident in droughts and floods that were disrupting Chinese agriculture. Leaders in Beijing were preparing extensive measures to slow greenhouse gas emissions. In the summer of 2007 they released a climate action plan mandating investments in renewable and alternative energy and improvements in energy efficiency. The plan proposed cutting about a gigaton of emissions from what otherwise would have been emitted. Similar plans were generated in India.

IN HIS LAST TWO YEARS IN OFFICE, Bush continued to soften his stand on climate change, largely in response to European pressure. At a G-8 meeting in Germany in June 2007, he endorsed the goal of a UN conference by 2009 that would produce an agreement on the part of the largest emitters and put the world on a path to cut in half total annual emissions by 2050. Bush joined this consensus at the behest of the meeting's host, German chancellor Angela Merkel, for whom climate change was a priority—and who was a staunch ally of Bush's in Afghanistan.

However, G-8 statements had the conspicuous defect of not reflecting the participation or consent of major emitters that were developing countries. A number of American experts on climate diplomacy had, for some time, been urging a separate, focused dialogue that would bring together representatives from the developed and developing worlds selected on the basis of their responsibility for the problem and their ability to contribute to a solution.[38]

In September 2007 Bush convened in Washington a group of sixteen countries—developed and developing—that represented nearly three-quarters of humanity, as well as 80 percent of all greenhouse gas emissions.*

In his speech at what was bluntly called the Major Emitters Forum, Bush said, "We acknowledge there is a problem, and . . . we commit ourselves to doing something about it. . . . We share a common responsibility: to reduce greenhouse gas emissions while keeping our economies growing."

While the administration continued to oppose Kyoto-style targets, it was now open to participating at the next UN climate conference, to be held that winter in Bali, Indonesia. The participants at that meeting were tasked with producing a "road map" for negotiating a successor to the Kyoto Protocol when it expired in 2012.

In Bali, a few small countries arrived with big promises that they hoped would inspire others to join them in making serious commitments. Norway and New Zealand, for example, pledged zero emissions by 2050. But the negotiators from the larger, more powerful delegations used Bali to protect their current positions and keep open their options for the future, particularly on what sort of binding targets should replace those agreed in Kyoto and how the Berlin Mandate would be applied to developing countries.

*Attendees included Australia, Britain, Brazil, Canada, China, France, Germany, India, Indonesia, Italy, Japan, South Korea, Mexico, Russia, South Africa, the United States, and the European Union, represented by Portugal, which held the rotating presidency. Representatives of the UN Framework Convention on Climate Change also participated.

They argued for two weeks. It was not until after the scheduled end of the meeting that the Indian delegation suggested compromise language for a final document: developing countries would undertake "nationally appropriate mitigation action." Those four words represented a lot of work as well as what diplomats call constructive ambiguity—that is, a deliberate lack of clarity or specificity for the sake of agreement. Developing countries were finally on record indicating that they would start cutting their own emissions. But they were careful not to link their pledge to any international negotiation or agreement. They also accepted language assuring that their reductions would be "measurable, reportable, and verifiable." But that promise was contingent on financing from the developed world and avoided any hint that there might be *international* verification of their compliance with their own targets.

The negotiators set a two-year deadline for resolving these issues, with negotiations to culminate at the next major meeting, which would be held in Copenhagen in December 2009—a year into the presidency of Bush's successor. The Bali Action Plan was sealed when Paula Dobriansky, the chief negotiator for the Bush administration, agreed to its terms.

DURING THE YEARS when Bush was largely boycotting the diplomacy of global warming, the U.S. Congress was quiescent on the subject. But now that he had put the executive branch back into the post-Kyoto process, concerns reemerged in the legislative branch about the potential cost to the American economy of reducing CO_2, especially if it involved an international treaty.

In early 2008 ten Democratic senators signed a letter to their leader, Harry Reid of Nevada, and to Barbara Boxer of California, the head of the Environment and Public Works Committee.

While polite in tone, the message was clear: advocates for aggressive action on climate change must ensure that whatever measures they proposed would protect American jobs.

This warning was a mild reprise of the Byrd-Hagel resolution that had crippled the Clinton administration in Kyoto; it was also a preview—expressed by Democrats—of the congressional caveats that a president of their own party, Barack Obama, would soon have to contend with.

CHAPTER FOUR

COALITIONS OF THE WILLING

UNTIL NEAR THE END of his long run for the presidency, Barack Obama put energy and climate at the top of his list of domestic priorities, above health care and education. He promised to implement a comprehensive (or "economywide") cap-and-trade program, reduce greenhouse gas emissions 83 percent by 2050, and return the United States to world leadership on climate change.

In the final weeks of the campaign, the nation and the world were hit by the worst crisis in the financial system since the Great Depression. The collapse had a profound and somewhat paradoxical effect on the politics of global warming. On the one hand, the recession in national economies made it harder for governments to contemplate commitments that would raise costs of any kind, including those that would discourage reliance on fossil fuels. At the same time, in their eagerness to revive their economies, nations were open to opportunities for public investment in clean energy infrastructure. Since the transition to low-carbon energy would entail up-front costs and mid- to long-term gains, it was in some ways impeded and in other ways spurred by the crisis.

When Obama moved into the Oval Office on January 20, 2009, the U.S. economy was still on the brink of an abyss. But even as he and his administration put together the American Recovery and Reinvestment Act to combat the recession, they looked for ways to encourage clean energy programs. Once the stimulus package was passed, the administration urged Congress to move quickly and simultaneously on climate and health care bills.

Meanwhile, the White House ordered the Department of Energy to enforce efficiency standards for household appliances; launched a program to generate more electricity from wind, wave, and ocean currents; and promulgated new fuel efficiency standards for cars. The Environmental Protection Agency took advantage of a 2007 Supreme Court decision that gave it authority under the 1990 Clean Air Act to declare that CO_2 and five other greenhouse gases were pollutants that endangered public health and welfare. Obama was demonstrating that he would make maximum use of the executive branch's authorities to regulate emissions if legislative measures stalled on Capitol Hill.

For a while, it looked as though Congress would do its part. Just before the Fourth of July recess, the House of Representatives passed the American Clean Energy and Security Act, cosponsored by two Democratic legislators, Henry Waxman of California and Ed Markey of Massachusetts. The vote was 219-212. It was the first time either chamber of Congress had acted on climate change. But the margin of victory was ominously narrow. Almost forty Democratic members opposed the bill. Most were from states that relied heavily on coal or heavy manufacturing.

The bill passed only because it had the support of eight Republicans. While that was hardly a mass defection, it was a meaningful break from the GOP leadership's decision to oppose the administration on virtually all major bills. Republicans like Mike

Castle of Delaware and Mary Bono Mack of California broke ranks because they were from districts with active environmental constituents. Leonard Lance of New Jersey voted with the Democrats because he hoped his constituents would benefit from the creation of "green jobs." John McHugh of New York—who became Obama's secretary of the Army—felt that the nation stood to benefit from reduced dependence on imported oil. And it was no coincidence that Obama had defeated John McCain in seven out of the eight Republican members' districts.

The bill, which was nearly 1,500 pages long, envisioned a sweeping overhaul of the U.S. energy industry. At its core was an economywide cap-and-trade system for CO_2 and other greenhouse gases. Capping would begin in 2012, then reduce permissible emissions to a prescribed amount each year thereafter. The intended effect was to cut the annual U.S. emissions from 5.7 gigatons to about 5 gigatons by 2020, then to under 3 gigatons by 2035, and finally to about 1 gigaton by 2050. This final number would be an 83 percent reduction in U.S. CO_2 emissions, in keeping with what Obama had promised during the election campaign.

Every major energy supplier and consumer in the country—from electric power utilities to heavy manufacturing companies to gasoline dealers—would be responsible for meeting the caps as they were steadily lowered. As the level of permissible emissions went down, the price of permits would go up, which meant that the cost of using carbon-based energy would rise.

Especially with the economy in terrible trouble, members of Congress worried about how high those costs would go, so the House bill borrowed a page from the George H. W. Bush administration's Clean Air Act and included a provision that allowed major energy users to buy and sell emissions permits. Companies that cut their emissions quickly and cheaply would have excess permits that they could sell to firms that could not meet the caps.

The net effect would be a dramatic reduction in the cost of compliance, but there would still be costs associated with implementation of the program. The Congressional Budget Office estimated that the price tag every year over the coming decade would be about $22 billion, or about $175 per household.

While many economists and environmentalists considered those costs to be a bargain, manufacturers and labor unions worried that they would raise the price of American goods and drive jobs to countries in the developing world. To address this concern, the bill threatened to require emission permits (called "border adjustments") for imports from countries that were not making a serious effort to reduce their emissions. The administration warned that this provision might touch off a new wave of American protectionism, which would provoke other countries to retaliate. A trade war is particularly dangerous during a global recession, when vigorous international commerce is crucial to recovery.

The dispute, which pitted Democrats against one another, was another warning that, whatever legislation ended up being passed on energy and climate, it would have to be sold politically as protecting or creating American jobs.

PASSAGE OF ENERGY AND CLIMATE LEGISLATION was even more difficult on the other side of the Capitol. A narrow margin of the sort that allowed the bill to squeak by in the House was not possible in the Senate, where a sixty-vote supermajority was needed to avoid a filibuster. Besides, Max Baucus, the powerful head of the Senate Finance Committee, which shares jurisdiction over any legislation that involves fees or taxes, insisted on giving priority to health care reform. By the fall, that issue had created a stalemate so severe that pending business on many other issues, including climate, came to a halt.

Obama would be going to Copenhagen in December for a summit devoted to reducing the world's emissions having failed to enact legislation that would reduce America's own emissions. Like his three predecessors, he was already at a disadvantage on the international stage, especially vis-à-vis the Europeans.

THE EUROPEAN UNION NOW CONSISTED of twenty-seven member states with an aggregate population of 400 million, which put it ahead of the United States and behind only China and India. A collective GDP of nearly $15 trillion made the European economy the largest in the world; the European Commission and the EU member states together constituted the world's largest single donor to international development, contributing over $50 billion a year; and the euro had become the second largest reserve currency and the second most heavily traded currency in the world after the U.S. dollar. Those attributes made the European Union not just a chorus of like-minded nations but a single influential voice on what its members felt should be the outcome in Copenhagen.

For seventeen years, the Europeans had been insisting on a legally binding climate treaty that would unite developed and developing countries. For months at the beginning of 2009, the Danish environment minister, Connie Hedegaard, reiterated that goal, frequently adding that there was "no Plan B."

Her fellow Europeans did everything they could to set an example and create momentum that would support that goal. For several years, the European Union had been committed to reducing emissions by 20 percent from 1990 levels by 2020. Early in 2009 it upped the ante by pledging to cut emissions by 30 percent if other developed nations took on similarly aggressive plans to reduce greenhouse gases. The challenge was clearly aimed at the new president of the United States.

Norway, which is not an EU member but coordinates its policies with the Union on many issues, went further: it would reward success in Copenhagen with its own pledge of a 40 percent cut by 2020. The British government called for the creation of a global fund that would eventually make $100 billion a year available to poor nations to help preserve their forests and adapt to the effects of global warming on their economies and societies. (Adaptation *to* climate change—that is, coping with its consequences—is considered distinct from mitigation *of* climate change—that is, slowing it down and ultimately reversing it.)

The Europeans were hoping to create the atmosphere of a charity auction. In addition to trying to goad or embarrass the United States into acting, they were outbidding each other in hopes that developing countries would accept reduction obligations of their own, even if less onerous than those of the developed world.

VARIABLE GEOMETRY

Obama did not want to cede to the Europeans leadership on climate change, nor did he have confidence in the UN-run process. Therefore he was looking for new mechanisms for managing the diplomacy of global warming. He had inherited two from George W. Bush.

The panic in the autumn of 2008 had frightened the world's leaders, including Bush, into beginning the process of replacing the G-8—the self-appointed board of directors of the global economy—with a new, larger body that would more equitably and effectively reflect the distribution of power in the early twenty-first century. On November 14, 2008, ten days after Obama's victory,

Bush hosted the inaugural meeting of the Group of Twenty, or G-20, at the leaders' level in Washington.* The word "climate" appeared only once in the 3,500-world communiqué, on a list of global threats ("We remain committed to addressing other critical challenges such as energy security and climate change, food security, the rule of law, and the fight against terrorism, poverty and disease"). Nevertheless, since there was a clear intention to make the G-20 a permanent fixture on the international landscape, China, India, Brazil, and South Africa now had seats at an important table where climate change would be prominently on the agenda in the future.

From its inception, the G-20 seemed likely to figure in Obama's presidency more than the G-8.

In April 2009—three months into his presidency—Obama hosted a meeting in Washington of the sixteen developed and developing nations that Bush had hosted in September 2007, known as the Major Emitters Forum. Since then, the members had lobbied to change the name to the Major Economies Forum on Energy and Climate, which preserved the initials but sounded less like a cartel of polluters. The group was soon known to cognoscenti simply as the MEF, and its participants generally felt it was an improvement on larger gatherings in that it brought the major players together in a setting that allowed for real give-and-take.

*The G-20 members: Africa is represented by South Africa; Latin America by Argentina, Brazil, and Mexico; North America by Canada and the United States; East Asia by China, Japan, and South Korea; South Asia by India; Southeast Asia by Australia and Indonesia; "Western Asia" by Saudi Arabia and Turkey; Europe by France, Germany, Italy, Russia, the United Kingdom—and the European Union. The G-20 has never actually met at twenty, and probably never will, because it has been open to extra invitees, or in the case of the first and second meetings, "self-invitees" from Spain and the Netherlands. The United Nations, the International Monetary Fund, and the World Bank are often represented as well.

The MEF was an example of what theoreticians and practitioners of international relations call "variable geometry." Rather than resorting by default to an existing organization, governments have often found it better to get together on the basis of a shared stake in a problem and a capacity to do something about it.

This practice is similar to what has, in recent decades, become common in the military realm: the establishment of "coalitions of the willing" to deal with security threats when the UN is unwilling or unable to act. During the Clinton administration, a NATO-led coalition of the willing expelled Serbian forces from Kosovo, and George W. Bush assembled a similar ad hoc force (though without NATO's blessing) to overthrow Saddam Hussein in Iraq. In both cases, the UN stayed out of the military operation but reengaged to help with postconflict reconstruction and in other ways.

Not coincidentally, the membership of the MEF was largely the same as that of the G-20, reinforcing the impression that the G-8 would eventually expand to include developing countries—and perhaps sooner rather than later. That impression was reinforced when five key developing countries (Brazil, China, India, Mexico, and South Africa) were invited to attend portions of the annual G-8 meeting in L'Aquila, Italy, in July 2009 and when a gathering of the MEF followed soon after.

L'Aquila was Obama's first G-8. The change in administration meant that the United States, Europe, and Japan could cooperate to an extent that had not been possible during the Bush years. They proposed a formula for "differentiation" whereby the developed countries would have to cut their emissions by at least 80 percent by 2050, but the reduction for the world as a whole would be only 50 percent. This 80-50 breakdown was intended to give the developing countries extra time before they would have to start reducing their own emissions.

The developing countries accepted the concept of the formula but not the numbers 80-50. They were not ready to take on numerical commitments for their own eventual obligations, and they wanted deeper cuts by the developed countries in the near term.

Two months later, at a G-20 meeting in Pittsburgh—the first to be hosted by Obama—the impasse remained. The assembled leaders of the developed and developing countries pledged, with much fanfare, that Copenhagen would produce "an agreement [that] must include mitigation, adaptation, technology, and financing." What kind of agreement they did not say, nor did they explain what the verb *include* meant. At best it would mean *discuss*. More likely it would mean *debate*.

By now, officials of several European governments were actively lowering expectations for Copenhagen. The conference's host, Danish prime minister Lars Løkke Rasmussen, began to speak about a *politically* binding accord instead of a legally binding one.* He was distancing himself from his own environment minister. There was a Plan B after all, and whatever its details, or lack of them, it was going to be easier for the United States to accept if only because it represented a European retreat from a binding treaty.

TRIANGULAR DIPLOMACY

As the EU-led effort to break the bloc-to-bloc impasse sputtered, nations on both sides of the divide wanted to show that even though a robust global deal was not in prospect, they were committed to making progress on their own.

*A politically binding commitment is a good-faith pledge that a nation will comply with an international agreement through its domestic laws. If a nation falls short of a political pledge, it suffers embarrassment but not sanctions or other forms of punishment that are provided for in a legally binding agreement, usually enshrined in a treaty.

In the weeks after the L'Aquila meeting, Indonesia and Mexico unveiled plans to cut emissions and Brazil took steps to stop illegal deforestation. A welcome competition seemed to be emerging among the major developing countries—including, crucially, China and India.

For months, Obama had been dispatching cabinet-level missions to Beijing and New Delhi to press for more ambitious and concrete national commitments to reduce emissions in preparation for summit meetings he would have with the Chinese and Indian leaders in November ahead of the Copenhagen gathering.

During the George W. Bush administration, the United States had been an easy scapegoat for the difficulties besetting the post-Kyoto climate process. With Obama's ascendance to the presidency, it was harder to blame the United States, and neither China nor India wanted to be seen as the spoiler in Copenhagen. Both governments played up their plans to limit emissions even as they held firm against being part of an international regime that required them to do so. China and India began to publicize their clean energy initiatives; increase investments in public transportation and electric power grids; and establish laws requiring more use of renewable energy sources, such as solar and wind.

During the fall, triangular diplomacy among the U.S., China, and India intensified. Obama's back-to-back November summits—with President Hu Jintao in Beijing and Prime Minister Manmohan Singh in Washington—were carefully choreographed to emphasize the positive. Hu and Singh—whose governments were consulting closely with each other on energy and climate issues—kept to a minimum their guilt-tripping of America for its carbon profligacy.

At the end of November, with Copenhagen right around the corner, China and India separately announced that they would double the reduction of their economies' "carbon intensity"—the amount of carbon emitted per unit of economic output. (Earlier they had said they would reduce intensity by 20 percent; now they promised a 40 percent reduction.)

Chinese and Indian per capita carbon emissions are only a fraction of those of developed countries. But their economies are more carbon-intensive than richer nations because they tend to use less modern (hence less efficient) technologies than developed countries. If the Chinese and Indian economies continue to grow at their current pace, their national emissions will likely skyrocket unless they shift away from fossil fuels. Setting targets for carbon intensity would require them to make that shift. China and India were associating themselves more clearly than they had before with the proposition that business as usual was not a responsible course.

However, the targets they had in mind were purely national ones, not in any way tied to those of other countries. In addition, the Chinese and Indian targets were pegged only to 2020, not to the longer term of 2030 and 2050 that the developed nations were pushing. In effect China and India were saying: "We don't know enough about our economic growth beyond the next decade to make any long-term commitments." That made it all the harder to calculate what other countries would have to do to cut global annual emissions in half, from 30 to 15 gigatons, by 2050.

Furthermore, China and India were vague about how they would report progress in meeting their carbon-intensity targets and how the rest of the world could be confident that they were in fact meeting those targets. The United States, the European Union, and others insisted on greater specificity about how monitoring and verification would work—a demand that China and India rejected as a violation of their sovereignty.

More generally, Chinese and Indian officials were at pains to deny that they were conceding anything to the developed world. As though to drive that point home, just before Copenhagen their environment ministers, who had been meeting during the year with those of Brazil and South Africa, got together as a foursome again in Beijing. As a self-appointed leadership group on behalf of the developing world, they adopted the name BASIC, which made more sense as an assertion of the role they intended to play than as an acronym for the countries involved.

The formation of the BASIC group was a reminder that the most prominent of the developing countries intended to use their own version of variable geometry to strengthen their position in Copenhagen. They threatened to walk out of the conference if the developed nations insisted on binding targets for developing countries. "We will not exit in isolation," said Jairam Ramesh, the Indian energy and environment minister. "We will coordinate our exit if any of our nonnegotiable terms are violated. Our entry and exit will be collective."

Despite this preemptive bellicosity, it was significant that, with Copenhagen looming, the two largest members of the new group—and the governments of the two most populous nations on the planet—had gone further than ever before in signaling their willingness to do their share, as they alone defined it, in the mitigation of climate change.

CHAPTER FIVE

A USEFUL DISAPPOINTMENT

AS ENVIRONMENTAL ACTIVISTS LOOKED AHEAD to the Fifteenth Conference of the Parties to the UN Framework Convention on Climate Change ("COP 15"), some optimists among them called the Danish capital "Hopenhagen."* After the meeting was over, the city had a new nickname: "Brokenhagen."[39]

What actually happened in Copenhagen during the twelve days in the second and third weeks of December 2009 was a potential turning point. In dashing unrealistic hopes, the conference provided an incentive to fix, or at least improve, a diplomatic process that was, if not already hopelessly broken, then moving much too slowly to keep pace with the process of climate change itself.

DURING THE FIRST TEN DAYS of the conference, environment ministers and other senior negotiators slogged and bickered their way to a framework for protecting forests and a sketchy outline

*"COP 1," in 1995, produced the Berlin Mandate. Subsequent COPs were held in Geneva, (famously) Kyoto, Buenos Aires (twice), Bonn (twice), The Hague, Marrakech, New Delhi, Milan, Montreal, Nairobi, Bali, and Poznan.

of agreements under which the developed world would provide poor and vulnerable countries greater access to technology and funding for reducing their emissions as well as for adapting to climate change. The agreement on protecting forests was nearly completed, but the ones on technology and financing for mitigation and adaptation were little more than vague and virtuous-sounding statements of intent.

The more the delegates talked, the less confident they were about whether the conference would make progress on targets for 2050, on the operational meaning of "common but differentiated responsibilities," and on whether an overall agreement should be legally or just politically binding.

Even as they wrangled among themselves, the participants were following the news out of Washington, where Barack Obama was fighting what looked like a never-ending and none-too-promising battle on health care. That spelled trouble for his administration's hope for getting an energy and climate bill from Congress to back up his pledge to cut U.S. emissions 17 percent below 2005 levels by 2020 and 83 percent by 2050.*

U.S. negotiators made clear that, without a climate bill, their country would not enter into a legally binding international agreement at Copenhagen. This was a reversal of the long-standing practice of first negotiating a treaty, then fighting for its approval in the Senate. In fact, even if Congress did pass a bill setting U.S. targets, it was doubtful the Senate would produce the sixty-seven votes needed to ratify a treaty coming out of Copenhagen or some

*There was also suspense about whether Obama would even attend the summit. He had already made one trip to Copenhagen in October, hoping—in vain, as it turned out—to persuade the International Olympic Committee to award the 2016 games to Chicago. He did not commit to return to Copenhagen until less than a month before the opening of the conference.

future UN conference unless it imposed what the Senate regarded as sufficient obligations on developing countries.

The Chinese delegation—headed by Xie Zhenhua, the vice chairman of the National Development Reform Commission, the most powerful single agency of the Chinese government—was openly resentful about Obama's having used the November bilateral summit in Beijing to press Hu Jintao to subject China's voluntary pledge on emissions intensity to international review and verification. China lobbied hard for other developing countries to support its opposition to this American demand and to keep pressing for a treaty that would be binding on developed countries.

Two days before heads of state and government were to arrive, representatives from the developing countries staged a walkout in protest over what they felt was unfair pressure on them to cut emissions. They were demonstrating negative unity—they knew what they were against—but they did not agree among themselves on how a "common but differentiated" deal (or, as many commentators were now calling it, a "binding but asymmetric" one) would work.[40]

In fact, fissures were growing within the bloc of developing countries. The smaller, poorer ones felt marginalized and condescended to by the larger economically dynamic ones—especially the BASIC group of China, India, Brazil, and South Africa.

The suspicion was growing among developing countries that China—as the world's largest consumer of fossil fuels and largest emitter of greenhouse gases on an annual basis—was blocking progress on any agreement in which it would be called to account for its own responsibility for climate change. China encouraged this impression by rounding up support for its hard-line position from some of the world's largest oil and gas exporters: Bolivia, Iran, Saudi Arabia, Sudan, and Venezuela, as well as Cuba, a likely future player in the world energy market because of the discovery of oil and gas reserves off its northern coast.

Yet another informal grouping was formed to counter the Chinese and keep the chances of an agreement alive. It included Algeria, Bangladesh, Ethiopia, Grenada, and Lesotho—all countries that wanted help with adaptation. They argued that a political accord was better than none at all—and that big emitters like China needed to be part of it. Two Pacific Ocean nations, the Solomon Islands and Tuvalu, whose existence was threatened by rising sea levels, broke with China in the opposite direction: they wanted a universally binding treaty—one that would apply to both developed and developing countries.

There were signs of at least tactical differences within the BASIC group itself. Presidents Luiz Inácio Lula de Silva of Brazil and Jacob Zuma of South Africa were deeply engaged in the negotiations, with each attending some of the working-level meetings. Lula seemed to be leaning toward a political accord that might lock in measures to protect Brazil's forests. Getting that outcome would require giving in to U.S. insistence on accounting and verification standards of the sort China was resisting. South Africa was positioning itself as the guardian of poorer African nations that were looking for financing to help adapt to droughts and to protect forests.

In contrast, Premier Wen Jiabao of China and Prime Minister Manmohan Singh of India did not participate in nearly as many of the meetings where the negotiators were slogging away. And while China and India maintained their joint demand for binding commitments on the part of developed countries, each was wary that the other might cut a separate deal with the United States.

The UN and Danish officials responsible for managing the negotiations were overwhelmed. The executive director of the

UN's climate unit, Yvo de Boer of the Netherlands, was doing everything he could to keep the negotiations from bogging down further, while the Danish prime minister, Rasmussen, frantically tried to draft a document that would bridge some of the differences, then held back on circulating it lest he be criticized for pushing his own views.

EVEN ON THE PUBLIC RELATIONS FRONT, the fates seemed to be conspiring against the conference.

Ever since Rio, the United Nations had been able to count on two favorable sidebars to media coverage of climate conferences: admiration for the reliable and increasingly detailed scientific evidence in support of the need for a global deal, and cheerleading from unofficial observers from the environmental community.

No such luck in Copenhagen. Two fast-breaking news stories cast fresh doubt both on the urgency of the climate change issue and also on the credibility and the competence of the United Nations as the organization to deal with it.

A firestorm burst over the release of hacked e-mail exchanges among scientists at the Climatic Research Unit of Britain's University of East Anglia who had been contributing to the work of the Intergovernmental Panel on Climate Change. The messages, which were supposed to be private, were construed by critics as suggesting that, in their zeal to sound the alarm, the same scientists were suppressing a paper that they believed was methodologically flawed and that also supported a forecast less dire than their own.

The incident was a bracing lesson in how careful IPCC experts had to be in what they thought were confidential and informal settings. Even though the IPCC had set and, to a remarkable degree, met the highest standards of quality control, skeptics and deniers had a gotcha field day that put the panel and the United Nations itself on the defensive in the press and the public debate.[41]

Over two decades, the IPCC's findings had earned widespread respect largely because they were scrupulous in considering every possible explanation for warming. The caution that the panel had applied to its deliberations was what made its conclusions so compelling in support of an increasingly stark warning. The e-mail controversy revealed an instance of close-mindedness to dissenting opinions that was damaging to the panel's reputation and out of character with its track record.

The other distraction was the result of an underestimation by the United Nations and the Danes of how many observers to expect at the conference. They prepared for 15,000 people, and nearly 40,000 showed up. The numbers swelled throughout the week, peaking just as the heads of state and government were swooping in with their entourages. Legions of representatives of nongovernmental organizations and advocacy groups, who were already angry about the lack of progress inside the conference halls, ended up stranded in the cold outside.

THE ELEVENTH HOUR

Meanwhile, the leaders, including Obama, who arrived near the end of the talks essentially had two choices: they could join those who were already there in presiding over a debacle, or they could throw themselves into the negotiations and try to salvage some sort of deal. They chose the second.

The hectic and suspenseful last stage of the conference began with the United States and several other industrial countries significantly increasing their pledges of financial support for poor and vulnerable nations. Secretary of State Hillary Clinton announced that the United States would help mobilize financing, from public and private sources in the developed world, in the range of $100 billion each year. That offer seemed to spark some willingness among representatives of the developing countries to consider making emission pledges of their own.

The only way to get any real business done was in small meetings—sometimes tête-à-têtes between key leaders. Arranging those is a challenge at a G-8 or a G-20 summit. But when more than a hundred heads of state and government are gathered at a conference, getting the right people around the right table at the right time is especially difficult, and the chances increase that connections will be missed, messages mixed, and signals crossed. In Copenhagen, the problem was compounded by the technical, economic, and political complexity of the issues at hand and by the lack of expertise on the part of many leaders who found themselves deep in the details where the devil resides. It was essential that they depart from protocol to seize opportunities as they arose, but the improvisational character of the side meetings contributed to the confusion and the potential for diplomatic blowups.

The most widely publicized incident that illustrated the element of serendipity at the summit was also the most consequential.

OBAMA ARRIVED EARLY IN THE MORNING on December 18 and planned to return to Washington later that day. He quickly sensed the mixture of panic and paralysis in the air, including on the part of the Danes and the senior UN officials who were supposed to be orchestrating the conference. Secretary Clinton, who had come the day before, told the president that it was "the worst meeting I've been to since eighth-grade student council."

Obama met privately with Wen Jiabao to see if they could reconcile the two points on which their governments had been at loggerheads for months: whether the new agreement would include domestic commitments from all the major countries, and whether it would stipulate the need for international review and verification of pledges. The session seemed to produce some progress. Wen did not push Obama on a successor agreement to Kyoto that would be binding for the developed countries, even though his negotiators had been unrelenting on that issue.

After their session, Obama and Wen returned to the offices set aside for their private use.

Around 4 p.m. word reached Obama that other countries were balking at the outlines of an agreement that he and Wen had sketched out, and the conference was abuzz with a rumor that another key leader, Manmohan Singh, was on his way to the airport. Obama was due to leave Copenhagen two hours later. With a major winter storm bearing down on Washington, there was not much room for delay. The president asked for a follow-up meeting with Wen as soon as possible. The Chinese delegation stalled, then finally agreed to meet at 6 p.m. only to come back shortly afterward and ask for an extension until 7 p.m.—an hour after Air Force One was supposed to be airborne.

The president's aides learned that Wen was closeted with his BASIC counterparts—Lula of Brazil, Zuma of South Africa, and Singh (so much for the earlier rumor about his departure). Obama decided to invite himself to the meeting.

Senior U.S. officials made their separate ways to the Chinese delegation's headquarters, where the BASIC meeting was under way in a room at the end of a long corridor, with a gaggle of journalists milling around outside.

Secretary Clinton got there first and was intercepted by a surprised and nervous Chinese security guard who steered her to a waiting room. She started to follow him, then realized he was trying to keep her as far away as possible from where the action was. She turned and headed back in the other direction just as Obama and his staff arrived. The astonished reporters parted to let him pass, and he made straight for the door to the meeting room. The now-desperate security guard realized he had lost control of the situation and tried to keep the Americans from bursting in on his boss. A White House staff member managed to get between the guard and the president, who entered the room, jovially greeted the nonplussed heads of state and government, and took a seat next to Lula across from Wen.

In Washington, the unexpected appearance of the president at a meeting in progress is known as a "drop-by." Inevitably, though, when the media learned what had happened, they had fun describing Obama as a gate-crasher, rather like the couple that showed up uninvited at the state dinner Obama had given for Manmohan Singh three weeks earlier.

Wen quickly recovered his aplomb and decided to make the most of the occasion. He handed Obama a draft statement that the four leaders had been working on. Obama read it quickly, complimented their good work, pronounced the document generally acceptable, but said there were "a couple of points" that needed further discussion: a political rather than a binding agreement, and verification of adherence to national commitments on the part of developed and developing countries alike.

When Lula objected, Obama turned to him and said that if the BASIC leaders were not prepared to be flexible on a few words, then he had plenty to do back home and would head straight for the airport. Lula quickly relented, then sat silently as Obama negotiated with Wen and Singh—and occasionally with the environment ministers and other officials who had been deeply immersed in the fine points of the talks.

When the proceedings turned contentious or seemed headed for a breakdown, Wen intervened to suggest compromises that Obama could accept.

Manmohan Singh engaged with Obama but let Jairam Ramesh, his energy and environment minister, do most of the arguing. Ramesh did so with relish. He was aggressive, sometimes acerbic, but not strident.

It was Obama who came up with what turned out to be mutually acceptable language that would commit developed and developing countries to make parallel and symmetrical *politically* binding pledges to reduce their emissions, with the understanding that "nationally appropriate mitigation actions [pledged targets] will be subject to international measurement, reporting and verification" (the formulation used in the final text).

Obama asked one last time if the other four leaders were prepared to associate their governments with the proposed language and whether their national pledges would be subject to verification. Wen assented, but Xie Zhenhua objected. Wen ordered the Chinese interpreter not to translate what Xie had said, and he ignored a second interruption. Wen, like Obama, was determined to narrow the gap between China and the United States and, as a result, save the conference.

At the heart of the deal were compromises on the nature and verification of pledges, as well as the first formal endorsement by developed and developing countries that global warming had to be stopped before it reached 2°C (3.6°F), with an option to tighten that target to 1.5°C (2.7°F) after a review of the science in 2015.

It had taken an unscheduled, unscripted meeting of five leaders—the American president and the four BASIC heads of state and government—to break the deadlock of the conference.

Just before Obama left, Wen asked him to persuade the Europeans to endorse the agreement, including leaving open the question of when or even if the political agreement would become a legally binding treaty of the sort the Europeans strongly preferred. In a rushed meeting, Obama elicited the unenthusiastic assent of Angela Merkel, the chancellor of Germany, Nicolas Sarkozy, the president of France, and Gordon Brown, the prime minister of Great Britain. Having gotten more from the BASIC leaders than they had wanted to give, Obama had now, as his last act in Copenhagen, gotten the EU triumvirate to accept less than it had hoped for.

The president left two aides, Michael Froman of the White House staff and Todd Stern, the chief climate negotiator, behind to deal with representatives of nations that were not represented in the small meeting. There was considerable consternation among countries whose governments were still committed to a legally binding successor to Kyoto, as well as among oil-exporting nations who opposed any agreement. Some leaders were angry enough not to endorse the outcome. As a result, the final plenary session, with all delegations present, merely "took note" of what became known as the Copenhagen Accord.

Even with an eleventh-hour compromise that saved what would otherwise have been a spectacular failure, the UN system could not fully accept the outcome because the process that produced it violated the protocols of universal consensus.

A FORK IN THE ROAD

Obama was among those participants in Copenhagen who returned home feeling relief that at least they had something to show for their efforts mixed with foreboding about what lay ahead.

Obama's challenge in Washington was to get an energy and climate bill out of Congress so that he could go to the next summit, in Cancun at the end of 2010, in a stronger position than he had been in at Copenhagen. Given the continuing stand-off in Congress over health care at the beginning of 2010, the new year looked like it might be every bit as tough on the legislative front as the old one had been. It might well be even tougher since the Republicans were gearing up for a mid-term election campaign in which they would pummel the Democrats on health care, whether a bill passed or not, and on the administration's desire for what the GOP was already characterizing as "a national energy tax." Some in Obama's party, and even a few within his administration, urged him to drop or at least delay his efforts on climate. But Obama still considered it a domestic priority.

Obama's diplomatic challenge was at least as daunting, and it was related to the one he faced with Congress. Much as he had inherited from his predecessors a national economy that had systemic flaws and treacherous vulnerabilities, he had also inherited a twenty-year international process for dealing with climate change that was not fulfilling the objectives it had set for itself and that was in tension with the exigencies of American politics.

The UN-led quest for the holy grail of a binding treaty as the solution to global warming had now become part of the problem in three respects. First, the process of preparing for and convening the summit had simultaneously encouraged false hopes, stirred up old grievances, exacerbated tensions, and nearly produced gridlock. Second, it had consumed—not very efficiently—the political equivalent of several gigatons of governments' and leaders' energy. And third, the summit, like every UN meeting on climate change, had given advocates of a global binding treaty a platform to chastise the United States for its recalcitrance on that issue, which was no help to Obama in his efforts to manage Congress.

Insofar as Copenhagen had brought these sad facts home to the world—and to Obama—it was a useful disappointment.

According to several of Obama's aides, it pained him to see the UN's shortcomings so spectacularly on display. Like the first President Bush and Bill Clinton, and in contrast to George W. Bush, Obama believed deeply in multilateralism—the idea that decisions affecting the planet as a whole should reflect, as much as possible, the interests and concerns of all governments that were prepared to play by the rules.

But on an urgent issue like climate change, multilateralism needs to work in tandem with what Moisés Naím, then the editor of *Foreign Policy,* dubbed "minilateralism"—the process of reaching agreement among the smallest possible number of countries needed to have the largest possible impact on solving a particular problem.[42]

Even as Obama and officials in his administration had been increasing their reliance in 2009 on smaller groupings—the G-8, the G-20, the Major Economies Forum—they were careful to describe what they were doing as "supporting the good work" of the United Nations.

THERE WAS A LOT TO SUPPORT—and therefore to retain and strengthen going forward. The UN had decades of experience providing services and fostering cooperation in ways that do not threaten the sovereignty of member states. During the cold war, the United Nations was most successful in dealing with issues that fell into four categories: humanitarian relief and public health initiatives that had little or nothing to do with the ideological and geopolitical rivalry between the superpowers; the protection of the environment; regulation on the use and spread of nuclear materials (a goal that Washington and Moscow both supported);

and coordination, quality control, and dissemination of scientific research relevant to public policy.

In the first category, the UN's World Health Organization helped eradicate smallpox, a disease that had, in the twentieth century, killed twice as many people as both world wars, the Holocaust, and the Stalinist and Maoist purges combined.[43]

In the realm of environmental protection, the UN's signal and lasting achievement was the Montreal Protocol. Signed in 1987 and updated in subsequent years, the pact has successfully limited the emission of chlorofluorocarbon compounds, commonly called Freon, and other substances that had opened up a hole in the ozone layer of the atmosphere, depriving the earth and its inhabitants of protection from harmful ultraviolet rays.

In the third category—nonproliferation—the job of monitoring and enforcing compliance with the Non-Proliferation Treaty belonged to the International Atomic Energy Agency (IAEA), a UN agency founded in the 1950s. From its base in Vienna, the IAEA is charged with making sure that nuclear energy programs around the world are safeguarded against accidents, misuse, and theft.

Having a robust nonproliferation treaty and a strong nuclear oversight agency will be essential for the mitigation of climate change. In theory, all nations should have the right to use nuclear reactors to generate electric power. Indeed, they should be encouraged to do so, because no fossil fuels are burned. In practice, however, the unregulated availability of nuclear technology for supposedly peaceful purposes would risk quickly turning into a bonanza for traffic in the material and know-how for making bombs.

Reactors are already the world's most widely used carbon-free way of generating electric power. The IAEA's capacity and authority will need to be enhanced and expanded to keep up with the widely advocated "renaissance" in nuclear-generated power as an alternative to high-carbon sources.

As for scientific research, the United Nations has spawned many specialized agencies that have earned the respect of academic experts and governments alike. Prime examples are the World Meteorological Organization and the United Nations Environment Program, which set up the Intergovernmental Panel on Climate Change.

In all four areas, the United Nations has worked best when member states have been given a chance to choose, quite literally, to *co*operate with it—that is, when national governments have sent their own personnel, under their own flags, to operate alongside UN agencies and contingents from other countries. In this respect, the United Nations has served as a facilitator rather than as a global superagency—not to mention a global superstate, with authority that supersedes that of its members.

A fifth category of global activity—the promotion and regulation of international finance, aid, and commerce—has been principally the province of the International Monetary Fund, the World Bank, and the World Trade Organization.

Of these, trade is arguably the most complicated case. Nations have always been inconsistent and in varying degrees at odds with one another about what constitutes free trade. Every nation wants as much access as possible to foreign markets for its goods, but it also wants to protect its farmers and industries from cheap imports. If the global marketplace were unregulated, every country would feel free to decide for itself what constitutes unfair competition. The result would be frequent trade wars that would be bad for everyone.

Much as the NPT and the IAEA have helped maintain the nuclear peace, the General Agreement on Tariffs and Trade, or

GATT, founded in 1947, served a similar purpose in keeping what might be called the trade peace by setting rules, adjudicating disputes, and reducing tariffs, thereby stimulating economic growth across the globe. But unlike the IMF and the World Bank, which are primarily in the business of helping countries advance their economic interests, the GATT's authority to help countries coordinate in liberalizing their domestic trade laws raised concerns about national sovereignty from the outset.

Much like the evolution of a global arms-control and nonproliferation regime, the emergence of a global trade regime in the second half of the twentieth century was possible largely because it came about slowly. Diplomacy had to stay in sync with domestic politics, which in turn were heavily influenced by domestic economics.

With that reality on everyone's mind, the gradual imposition of international regulations did not outpace individual countries' willingness to be subject to those regulations. Governments, parliaments, and private sectors had time to build up confidence that they would benefit from membership in a trade regime; that they could trust other members to play by the rules; that the rules themselves were fair; and that the international authorities supervising the process were competent and evenhanded. Only then would nations take on obligations for the sake of the "global public good" of free and fair international trade. After nearly fifty years of trial and error, nations around the world finally had enough trust in the credibility and utility of the system to transform a "general agreement" into a global institution: the GATT became the World Trade Organization.

In that respect, the experience with trade provides a lesson and a model for negotiating complex international agreements that have an impact on sensitive domestic industrial sectors as well as for establishing worldwide markets for environmental goods and services, particularly in the field of energy.

WHERE THE UNITED NATIONS HAS COME UP SHORT has been in its effort to convene and guide the negotiation of an energy and climate treaty. Its failure has been, in large measure, the flip side of its virtue. The UN's universal membership and comprehensive agenda make it, in certain respects and on many issues, indispensable and irreplaceable. The smallest, poorest, and weakest countries have a forum where they can be heard along with the largest, strongest, and most prosperous. That capacious representation is particularly important on the issue of climate change since small and poor nations are among the most vulnerable to desertification and flooding. Yet the UN is too large, too inclusive, and too limited in its authority to move quickly and decisively.

In *The Parliament of Man,* published in 2006, the historian Paul Kennedy credited the world organization for giving substance to the idea of an "international civil society."[44] That is no small thing. Civil society is a prerequisite for good governance; but it is not, by definition, a governing body or, in and of itself, a system of governance. It is not at the United Nations that the decisions to limit and then reduce greenhouse gas emissions will be made. Rather, those decisions will be made in the capitals of the organization's member states. The General Assembly is still a long way from being the parliament of man. It is in national legislatures that treaties will be approved and given the force of law—or, as sometimes happens, rejected or ignored.

THAT SOMBER SET OF FACTS about the limits of multilateralism was on prominent display in 2009. The compensatory good news was that the year had seen a boom in minilateralism: the reconvening of the Major Economies Forum and increasing attention to climate in the G-8 and the G-20. The positive dynamic that had developed during the run-up to Copenhagen—when individual countries tried to outdo each other with forward-leaning pledges—showed that competitive unilateralism has its own benefits.

As for the Copenhagen conference itself, Obama's drop-by on the BASIC leaders' meeting and the quick huddle he held afterward with the EU leaders broke the deadlock and produced the best deal possible under the circumstances. It was far from a breakthrough, but better than breakdown—and more than a placeholder, in that it established an agreement on the red line that must not be crossed on the planet's fever chart as well as the core principle of verifiable national limits, even if they were only politically binding and lacked the UN's formal approval.

Thus, the lesson of 2009 for 2010 and beyond is that a small number of key countries should concentrate on coordinating their national policies as much as their domestic politics will permit. As that process continues, it might create its own gravitational pull; the right inner circle can expand outward, bringing other countries into a virtual global compact that will take effect much more quickly than a legally binding treaty can be negotiated—not to mention ratified by the U.S. Senate.

CHAPTER SIX

THE BIG FOUR

THE INNER CIRCLE FOR THE NEXT PHASE of climate diplomacy already exists. It is made up of the United States, the European Union, China, and India. They account for nearly half the world's population (3.3 billion out of 6.8 billion), 63 percent of the global GDP, and two-thirds of all civilian nuclear reactors. The four are also the world's top four emitters of carbon dioxide, accounting for about 60 percent of the total.

Moreover, the Big Four bridge the divide consecrated by the Berlin Mandate. The United States and the European Union clearly lead the developed world, with a combined population of over 800 million people, a combined GDP of $29 trillion (about 42 percent of the world's $70 trillion economy), and combined annual emissions of about 9 gigatons of CO_2 (about 30 percent of the world's total). At current rates of growth, their aggregate population will be well over 1 billion by 2050, and their aggregate economy will have doubled.

If the United States and the European Union can pass the laws and make the changes in their economies necessary to cut emissions by over 80 percent, their combined annual emissions in

2050 will be just 1.5 gigatons, a level that would help the world stay in the range of 15 gigatons and keep global temperatures from rising into the danger zone.

China and India occupy a place of similar predominance in the developing world. Their combined population is 2.5 billion (more than a third of the world's total), and their combined GDP is $11.2 trillion (a sixth of the world's). Together, they produce more than a quarter of the world's CO_2 emissions (7.6 gigatons).

In Copenhagen they made clear that while they expect their emissions to grow along with their economies, eventually their emissions will peak and then start to decline—although later than will be the case in developed countries. The big question for China and India is whether the peak will come soon enough—and whether the subsequent decline will be sharp enough—for the world to make its 2050 target.

It is noteworthy that of the four, the one with the most impressive record is not even a state. Yet on the issue of climate change, the European Union not only acts as though it were a single state—so far it has acted more effectively than any of the other three major players.

For twenty years, the European Union has made cutting emissions a priority. It has used regular meetings of leaders and ministers to agree on common goals and policies, taking advantage of opportunities when they saw them, including learning from American innovations. In 2005 the Europeans borrowed from George H. W. Bush's cap-and-trade system for reducing sulfur dioxide and used it to impose mandatory reduction of CO_2 under the European Emissions Trading System.

The rest of the world is a long way from agreeing to a single price on carbon and having a global market for emissions trading.

That makes it all the more useful to have, in Europe, the first system for putting a price on carbon. After getting over some technical hurdles in its trial phase, the Emissions Trading System is now working reasonably well. Not only do environmentalists endorse it, but so do corporate leaders, because the predictability of increasing costs of carbon over the next forty years allows European companies to plan their capital investments.

In the best possible way, history is repeating itself. Just as the European Coal and Steel Community established the core of the Common Market, which in turn led to the EU itself and an expanding zone of peace, the Emissions Trading System is laying the basis for a much larger—perhaps ultimately global—transnational carbon market and zone of climate management.

The European Union is a model for the world in another respect as well. More than a decade ago the Europeans were able to go to Kyoto with a unified position that contained a workable regional analogue to the Rio treaty's leniency to developing countries. The EU put in place a "common" set of goals that "differentiated" between wealthy and poorer EU member states. Those in the first category accepted aggressive emissions targets, while those in the second were at first required only to slow the growth of their emissions before ultimately having to cut them.

Phasing out those disparities over time will be a difficult test for the European Union. As the current monetary crisis within Europe has proven, it is relatively easy for the member states to agree to common goals but much harder to rein in countries that do not make their targets. The biggest challenge lies in the poorer countries of the south and those in the east that are emerging from a communist past, where it is still especially hard to reconcile environmental protection with economic growth.[45]

Furthermore, Europeans should not forget that their leadership on this issue came about largely because of advantages derived from their system of proportional representation and the coincidence of three big countries slashing their emissions in the early 1990s for reasons having nothing to do with climate change. Fulfilling the EU vision of a truly unified system that pools national sovereignty on energy policy will require the more prosperous and "greener" states of northern Europe to subsidize clean energy development in the south.

Spokesmen for the northern member states have often touted the EU climate compact as the precedent for a global treaty in the future. Perhaps. But for some time to come the EU will have its hands full sustaining, solidifying, and paying for its own promising but difficult work in progress.

THE SITUATION in the United States is nearly the opposite of that in Europe: for two decades, America has disappointed the world and many of its own citizens in being slow to get its act together on climate; it has yet to agree on its goals and the means to attain them. But once federal laws are on the books, they are likely to be enforceable in a way that the European Union will envy—and perhaps emulate.

The trio of senators (a Democrat, a Republican, and an Independent) who joined forces in the spring of 2010—John Kerry, Lindsey Graham, and Joseph Lieberman—all endorsed the House-approved CO_2 reductions of 17 percent by 2020 and 83 percent by 2050, but the new Senate bill would erect a regulatory system more slowly, applying cap-and-trade only to electric power utilities, at least in the first few years. The bill would cover industrial emissions by a straightforward cap and transportation fuels with a fixed fee (which opponents are sure to describe as a tax). It would also include more support for nuclear and offshore oil drilling than the House bill—provisions that could be critical to gaining the support of some Republicans (including Graham himself, who threatened in late April to drop his cosponsorship of the bill because he sensed that Senate Democratic leaders would not make it a high enough priority).

Another bipartisan pairing—Maria Cantwell, a Democrat from the state of Washington, and Susan Collins, a Republican from Maine—introduced their own bill with many of the same objectives of Senators Kerry, Graham, and Lieberman but with a more cautious approach to emission trading.

Getting to the magic number of sixty votes for either bill, or a combination of the two that might include additional sponsors, will be an uphill battle, requiring the support of more than two dozen moderate Democrats from coal states and Republicans who favor investments in nuclear and renewable energy. A number of industrial-state moderates have expressed tentative willingness to join a bipartisan effort.

So the best to hope for in 2010 is an outcome that begins to send a "price signal" to major energy producers and consumers that the cost of emissions is going to go steadily up. The more detailed the signal is, and the more warning it provides about what to expect in the future, the better. Ideally, any legislation that passed would establish a schedule for targets stretching over the coming decades, so that energy producers and consumers can start making near-term investments and adjustments as part of their long-term planning for a low-carbon future.

If all this can be accomplished by the fall, the United States and the European Union might be able to demonstrate at the upcoming climate summit in Cancun—and at subsequent meetings—that a politically binding agreement is worth quite a bit and can be the basis for genuine cooperation. That would help convince India and China that they too should follow through on their Copenhagen pledges. It is essential for them to do so, because even the most heroic achievements by the Big Two of the developed world will be insufficient unless the Big Two developing countries do their part.

THE ASIAN GIANTS

The emergence of China and India as major actors on the world stage coincided remarkably with the history of climate negotiations. The Rio Earth Summit of 1992 came soon after China's economic liberalization and at the same time as India's.

China's reforms—which are based on Deng Xiaoping's "Four Modernizations" (all in energy-relevant fields: agriculture, industry, national defense, and science and technology)—put the country on course toward becoming what World Bank president Robert Zoellick has termed a "responsible stakeholder" in a rule-based world order.

India's prime minister, Manmohan Singh, was, at the time of Rio, the finance minister who spearheaded the opening of his country's economy to the world. As prime minister since 2004, he has nudged his government toward a more constructive position on climate change. That is good news since a climate management regime will need India cooperating—or even competing with—China for responsible leadership of the developing world. Depending on how fast their economies, populations, and energy usage increase, China and India could, by themselves, put as many gigatons of CO_2 and other greenhouse gases into the atmosphere as the developed countries are now pledging to cut. China and India could even exceed the 15-gigaton goal for the whole world in 2050. In fact, even if one of the two giants participated fully and the other stayed out, the very difficult task of bringing global annual CO_2 emissions down to 15 gigatons by 2050 would be unattainable.

India's economic growth over the next two decades could increase its energy needs fivefold. As agriculture moves from subsistence farming to modern industrial agriculture, it will become increasingly mechanized. That will lead to greater demand for oil and gas, as well as for electricity to heat homes and drive irrigation systems. As Indians migrate from the countryside to the cities to fill jobs in the manufacturing and service sectors, urban transportation infrastructure will have to expand, and more apartment complexes, factories, and office buildings that require air-conditioning will have to be built.

Given all these upward pressures on India's energy needs, its CO_2 emissions could double from about 1.5 gigatons to about 3 gigatons over the next decade. By 2030 emissions could rise to anywhere from 4 to 6 gigatons. That would approach or exceed one-third of the world's total allotment.[46]

China's annual CO_2 emissions, which are already the highest in the world, will only increase in the coming decades. Under some projections, China's emissions could double by 2020, taking it to more than 12 gigatons by then—just when the world as a whole needs to be bending the curve downward.

Fortunately, the argument against letting this happen is compelling for India and China themselves. Unless they can find a way of getting their own emissions under control, they will be as much at risk to the consequences of climate change as the rest of the world—and in some ways, even more so.

Several studies suggest that more Indians are directly threatened by climate change than any other people on earth.[47] The disruption of summer monsoons could severely affect the hundreds of millions of Indian farmers who would be at the mercy of droughts and floods. An increase in suicides by impoverished farmers has been attributed to heat-induced crop failures—a real-world tragedy that is beginning to make the possible impact of climate change a subject of interest in the Indian media and a concern of parliamentarians for their constituents.

Comparable numbers of Indians live near the 4,000-mile-long coast, including on the outskirts of megacities like Calcutta, Mumbai, and Chennai. The 1999 "super-cyclone" in Orissa state, which caused 30,000 deaths, and the 2006 Aceh tsunami, which killed about 8,000 in India and drove 140,000 from their homes, were warnings of how much at risk India's population is to coastal storms and the rise in sea levels that is already occurring as a result of climate change.

China is not far behind in its own vulnerability. That was the conclusion of a Chinese government report released in October 2008.[48] China's low-lying eastern provinces are home to over 90 percent of the population and therefore, like coastal India, threatened by rising sea levels and severe storms; and its northern and western provinces have been through a decade of devastating droughts.

In addition, both India and China suffer from mounting air pollution. Soot, or black carbon, is a major cause of disease. But it is also a serious contributor to global warming.[49] So by focusing on it as a public health issue, India and China will be escalating their fight against climate change. If not addressed, black-carbon-related warming could hasten the melting of the Himalayan glaciers, the world's third largest ice cap. The added runoff of winter snows feeds the Yangtze River in China and the Ganges and Brahmaputra in India. That could be a contributing factor to greater flooding in both countries. What exactly is happening and what will happen in the future are not well understood. But policymakers, including India's Jairam Ramesh, are increasingly concerned that any disruption in the timing and rate of runoffs from the glaciers could whipsaw cities and rural areas with droughts and floods.[50]

Recognition of these grim realities—the good news of damnation—figured in the Indian and Chinese decisions, before and during the Copenhagen conference, to shift, albeit cautiously and with multiple disclaimers, toward unilateral reduction measures that might fit usefully into an informal multilateral compact.

As Prime Minister Singh charts India's energy future, he must avoid several related pitfalls that he is reminded of every day—by the press, by opposition spokesmen in parliament, and by his own ministers. He must not appear to acquiesce to the dictates of the developed countries on climate issues; he must avoid agreements that can be depicted as discriminatory; and he must not appear to be applying the brakes to the juggernaut of the Indian economic boom, especially if doing so will deprive the world's largest underclass of a chance to join the world's second largest middle class.

Indians have for years made much of the difference between their own per capita carbon footprint and that of the United States. The average American is responsible for emitting fourteen times as much CO_2 as the average Indian. That stunning discrepancy is partly a result of the interplay of differences in levels of economic development, government policy, population trends, and energy use.

The population of the United States is a quarter of India's, but its CO_2 emissions are four times India's and its per capita GDP is fifteen times greater. During its first forty years after gaining independence in 1947, India was caught in a vicious cycle: poverty tended to promote large families, especially in rural areas, putting population growth on a steep upward curve that outpaced China's. (That was largely because of a strictly enforced one-child-per-family policy adopted in 1978.) India is now on a trajectory to pass China as the most populous nation around the middle of the century, when each nation will near or surpass the 1.5 billion mark.

India's population growth will steepen the national emissions curve disastrously if the economy continues to grow and remains dependent on fossil fuels. India's challenge is to reverse the vicious cycle. It has the potential to do so. As a highly entrepreneurial society that has already made considerable headway in developing a knowledge-based and high-tech economy, India may well be able—as Singh hopes—to lower birthrates by raising urban and rural living standards while developing low- and no-carbon sources of energy.

While fending off charges that he is under pressure from the outside world, Singh is making an explicit economic case for transitioning to greater energy efficiency. It is possible—and some authoritative Indians have hinted as much—that India may go beyond reductions in carbon intensity and commit itself to substantial reductions in emissions down the road—but only if developed countries first make good on their own pledges.

Some Indian experts believe that over the next two decades, their country might be able to cut its annual CO_2 emissions in half—from the 4-to-6 gigaton range where they are now headed, to the 2-to-3 gigaton range by 2030.[51] There is serious consideration of a truly comprehensive climate action plan. Its goal, over the longer term, is to reduce India's reliance on imported coal, oil, and gas, and in the near and medium term to reduce black carbon and invest in renewable energy sources: solar, hydroelectric, and wind.[52]

FAR MORE CONTROVERSIAL is nuclear energy, which will be critical for India as it will be for much of the rest of the world. A number of Indians have been drawn to the cause of climate change precisely because it strengthens their argument for accelerating the development of nuclear power. India has announced a highly ambitious plan to build twenty-one nuclear power plants in the next few years, with a combined capacity to produce three times as much electricity as Los Angeles uses today.[53]

The United States, the European Union, and others can help India with these and other plans, but it should be in the context of India's doing more to stop the spread of nuclear weaponry. India is one of the few countries that have never joined the Nuclear Non-Proliferation Treaty, making itself ineligible for international assistance in developing peaceful nuclear energy. When India conducted a series of tests in 1998, it was slapped with sanctions by much of the rest of the world.

The George W. Bush administration not only lifted sanctions that were still in effect when he entered office but granted India a cost-free exemption from the NPT, giving it virtually all the benefits enjoyed by non-nuclear-weapon states. India could somewhat reduce the danger of other countries breaking out of the NPT if it signed a treaty banning the testing of nuclear weapons, joined the United States in pushing for a prohibition on the production of fissile material that can be used in atomic bombs, and took various measures to avoid an arms race with its nuclear-armed neighbors Pakistan and China.

So far, India has refused to concede, even in principle, that there is a nexus between its own need for nuclear energy as an alternative to fossil fuels and the world's need for a truly universal global nonproliferation regime. However, Indian officials have hinted that they might take a step in the right direction by joining the Comprehensive Test Ban Treaty if the U.S. Senate ratifies it later this year. (The Senate rejected the Comprehensive Test Ban Treaty in 1999. Obama has vowed to resubmit it for ratification, although it faces strong Republican opposition.)

Thus, as India continues its emergence as a major power, there are two ways it can serve both the world and itself: an energy policy that permits economic growth while putting India on a path toward emission reductions; and a nonproliferation policy that will allow it to get maximum benefit from peaceful nuclear energy. The more willing India is to help on both those causes—and they are related, since nuclear energy is an alternative to fossil fuels—the more the United States and the European Union will be able to help India.

For its part, China has already embarked on a deliberate, serious, but cautious campaign to develop a postcarbon economy—and not a moment too soon.

The breakneck speed of China's economic development demands enormous energy, more than 70 percent of which now comes from coal. Coal plants in operation today are intended to be in service by midcentury. If, according to current plans, the Chinese government moves 400 million people from rural areas into urban ones over the next three decades, the cities' energy needs will rise accordingly. And now that many couples are allowed to have two children, the population rate may grow after decades of having held steady.

With the implications of all this in mind, China has redoubled efforts to improve the efficiency of factories, homes, and office buildings. It has adopted a more stringent standard for automobile fuel efficiency than the United States, and it is moving more rapidly than the United States and Japan toward developing next-generation electric vehicles. China already has become a major producer of wind power and is primed to become a leading exporter of both solar and wind technology.

This cluster of programs is as expensive as it is necessary. McKinsey & Company, which has worked closely with the Chinese government and companies, estimates that the measures necessary to keep China's emissions to 7 gigatons by 2030 will require an investment of $4 trillion to $5 trillion over the next twenty years. Even if that cost can be shaved through improvements in efficiency, China's energy transformation will require over $1 trillion in up-front investment.

As with India, the United States and the European Union can be vital partners. American and European firms are working with the Chinese on relatively inexpensive ways to cut black carbon emissions and cooperating in the development of the next generation of green technologies, particularly the capture and storage of CO_2 from power plants and the expanded use of renewables.

CHINA CAN ALSO BENEFIT from coordination with the United States and the European Union in the evolution of its system of governance more generally.

It has become almost conventional wisdom to believe that China, as a top-down, meritocratic, authoritarian state, has a competitive advantage over democracies. There is no question that long-range planning and disciplined, methodical execution of government plans have paid off dramatically in an ability to modernize infrastructure with a speed and on a scale that would have dazzled Deng Xiaoping. The Chinese government's capacity for making decisions and setting strategic priorities has often led Westerners to lament the contrast to the circus-like spectacle of their own political system in action—or, as was often the case in Washington over the past year, inaction.

In fact, though, the Chinese, like the rest of the world, are making it up as they go along, grappling with how to reconcile environmental and economic self-interest. Speaking privately, quite a few knowledgeable Chinese recognize that being an export powerhouse in the field of green technology is not going to solve their own climate and other environmental challenges. Their determination to beat Ford and Toyota in the development of electric cars does not mean they have a grand plan or a consensus within the leadership on how to make the transition to energy efficiency and low-carbon economic growth.

Furthermore, China is still a long way from figuring out how to encourage adaptation to, and enforce compliance with, the targets its leadership is now beginning to set, starting with the energy-intensity goals it announced on the eve of Copenhagen. The problem, in essence, is that the Chinese system is not currently able to translate targets into practice. As two Brookings scholars—Kenneth Lieberthal and David Sandalow (who is now an assistant secretary of energy)—have pointed out, "Despite increasing efforts to build environmental and energy efficiency concerns into the incentive structure all the way down the line, most local officials still regard meeting GDP growth as their primary objective."[54]

Safety is an issue, too—both for the population and the environment. As China proceeds with its own full-speed nuclear renaissance, it is going to have to train, mobilize, and deploy an army of inspectors.[55] By reliable accounts, only 10 percent of China's environmental laws have actually been implemented.[56]

These problems will persist as long as China's economic and managerial culture are geared almost exclusively to growth, virtually for its own sake.

In that context, it was encouraging that Wen Jiabao supported Obama's insistence in Copenhagen that China provide reliable and verifiable information about its progress in meeting the goals that it is now setting for itself. Not only did Wen overrule dissent within his own negotiating team—he stood up to a firestorm of criticism on his return to Beijing. The flak came from two quarters. Progrowth hardliners assailed him for making too many concessions, while environmental experts and activists complained that he had not gone far enough to protect the planet.

That debate echoes the American one—except in China, almost no critic of the government wants to be quoted in the press. Nearly two months after Copenhagen, the Politburo decided, in the words of yet another unnamed source, "to stop the debate and force everyone to focus on the implementation, rather than the virtue, of the decision" to stick with Wen's commitments under the Copenhagen Accord.[57]

It is a safe guess that Wen was not just giving in to Obama for the sake of a deal. He probably understood that he and other reform-minded leaders can use the give-and-take with the United States and other countries to force changes in the Chinese system so that it can bring the carbon-belching engine of its economy under control—and, over time, swap it out for an industrial system that is as energy efficient as posthybrid autos.

WIDENING THE CIRCLE

While concentrating on India and China, the United States and the European Union should step up their outreach to the other

BASICs—Brazil and South Africa—as well as Japan (the fourth largest economy in the world after the EU, the U.S., and China) and Russia (which has the most oil and gas reserves outside the Middle East and a fifth of the forested land on the planet). The resulting group—yet another G-8—would account for three-quarters of the world's emissions.

The next concentric circle might include the remaining members of the Major Economies Forum—Australia, Canada, Indonesia, Mexico, and South Korea as well as the four major member states of the European Union: France, Germany, Italy, and the United Kingdom. All those nations' leaders are now meeting on a regular basis as the G-20, where they are joined by Argentina, Turkey, and Saudi Arabia. Working with that group, the Big Four can accelerate cooperation on measures that will lead to a brighter, cleaner energy future—and, just as important, head off or tamp down disputes within their own ranks that might otherwise impede progress.

As some of these countries put a price on carbon and create mechanisms to trade emission permits within their own borders, they might start trading internationally as well. Cross-border trading is already happening within the European Union. A global carbon trading zone was envisioned at Kyoto, but nothing came of it, in part because it would have to be established and enforced by a legally binding treaty. That should not be necessary for an emissions trading system, just as it was not necessary for the General Agreement on Tariffs and Trade.

In contrast to the Kyoto Protocol, which would have subordinated a state's policies to the decisions of an international organization, an approach that might be called a General Agreement to Reduce Emissions (GARE) would perform the GATT-like function of setting rules, arbitrating disputes, and creating incentives for still other countries to coordinate in reducing emissions.

Nations could join this club by adopting their own ambitious and verifiable reduction targets grounded in domestic legislation. So, while the international dimension of the emerging global compact would be "only" politically binding, it would be anchored in legally binding national obligations.

Countries would work together to make sure that all members of the GARE have reliable reporting, monitoring, and enforcement systems in place—a prospect that Obama and Wen Jiabao made more likely with their agreement during their impromptu final meeting in Copenhagen.

Once participating countries' laws are sufficiently ambitious in reducing emissions and once they have confidence in one another's compliance with their own targets, international emissions trading is the logical next step. As in the GATT-nurtured free trade regime, a single set of rules would lower the transaction costs for participants; investors would be more likely to fund projects in countries with the most cost-effective emission-reduction policies; and that would be an incentive for *all* countries to tighten their own controls.[58]

ANOTHER VIRTUE of the GARE would be particularly pertinent to the United States, with its high barriers to legislative approval of treaties. Because the GARE would not be a treaty but an agreement, it would require a sixty-vote majority in the U.S. Senate (because of the controversial supermajority required to prevent a filibuster), but at least it would not need the sixty-seven votes required for ratification of a treaty.

Indeed, most of the provisions needed to validate emissions trading with other countries could actually be built into U.S.

domestic legislation already under consideration in Congress. Therefore no additional legislation would be required, because the House-passed bill—in a provision likely to be considered by the Senate—already authorizes the EPA administrator, in consultation with the secretary of state, to trade emissions permits with any "national or supranational foreign government [that is, the EU]" that imposes a mandatory cap on greenhouse gas emissions, either across the whole economy or across a major sector. This provision also requires the Environmental Protection Agency to determine that the foreign country's program is "at least as stringent as the program established by this title, including provisions to ensure at least comparable monitoring, compliance, enforcement."[59] In short, reciprocal domestic laws obviate the need for a treaty.

THE BUMPER STICKER for GARE on American cars (gas-guzzlers as well as hybrids) might be: "legislate nationally, coordinate globally."

U.S. foreign and national security policy has been hobbled for decades by the executive branch's habit of first negotiating with foreign governments, then bringing a treaty back home and trying to sell it to legislators. As a result, treaties have been susceptible to second-guessing; the attachment of troublesome, if not "killer," amendments; or death in the ratification process.

If bipartisan Senate legislation survives the current debate and becomes law by the early fall, it will not only strengthen Obama's hand at the Cancun climate conference—it might also give him a chance to explore the idea of a GARE in future smaller meetings, such as the G-20 scheduled for November in Seoul.

After all, the leaders gathered under the aegis of the G-20 represent 66 percent of the world's population and their countries produce 84 percent of the world's emissions.[60] That is not as inclusive as it ideally should be, since every nation—and, for that matter, every person—on earth should be a responsible stakeholder in climate management. But the only piece of variable geometry that accommodates all humanity is the G-192, better known as the United Nations, which still has an important role to play in many areas, but not as the principal catalyst for meaningful progress toward an effective global deal in the years immediately ahead.

CHAPTER SEVEN

THE SHADOW OF THE FUTURE

BEFORE PRESIDENT OBAMA CAN SUCCEED as a world leader on the issue of global warming, he has to succeed as a national leader. That means working with both parties in both houses of Congress to pass comprehensive climate and energy legislation. For that domestic effort to have maximum benefit diplomatically, legislation should be signed into law this year, before the Cancun climate change summit in November and December.

As Obama throws himself into that task, none of his political skills will be more important than his ability to use words as a political tool. His strength in this respect was evident from the beginning of his presidential campaign in 2007 to his rally with House Democrats just before the passage of the health care reform bill in late March 2010.

Making the case for a bill that will be as controversial as it is essential, he might draw from some precepts of the Golden Age of Greece, two and a half millennia ago, that are remarkably germane to the ethical and political challenges that come with the Age of Global Warming. Turning to that source of wisdom should come naturally to the first American president to have identified himself as a "citizen of the world" even before he was elected—and on foreign soil, no less—thus associating himself with Socrates. That much-noted line in Obama's Berlin speech of July 2008 burnished his credentials as a liberal internationalist—and, in the eyes of the right, heightened his vulnerability to the charge of being a one-worlder, or worse, not a real American.[61]

But it is Aristotle's philosophy that most resonates with Obama's political instincts and the test to which they will be put during the U.S. domestic debate over climate change. Aristotle's thinking about the nature of leadership in a democracy emphasized the importance of pragmatism, reason, and civil political discourse of the sort that Obama brought to his run for the White House and tried to bring to the debate over health care. Aristotle was also a theoretician and connoisseur of rhetoric—an attribute of Obama's for which he has been both praised and scorned. But for Aristotle, rhetoric meant more than eloquence; it was a strategy for a winning argument that depended on what he called "the three artistic proofs," or means of persuasion: logos, ethos, and pathos.[62]

Logos means command of facts, analysis, and responsible, carefully qualified guesses about the future, along with the ability to marshal them rigorously (and logically) on behalf of sound policy.

The Intergovernmental Panel on Climate Change is in the logos business. Its assessments and forecasts have been based on intense scrutiny, caution, and a willingness to be forthright about what is known and what is not known about the climate.

The recent controversies—sometimes collectively referred to as "Climategate"—have been exploited effectively by deniers and skeptics. They have had success in attacking the premise of global warming. Now, a majority of those who accept the existence of climate change believes it is attributable mostly to natural causes. And a Gallup poll in March 2010 found that 48 percent of respondents suspect that the threat of global warming is "generally exaggerated"—an increase from 35 percent two years ago.

This setback to America's collective logos increases the onus on the scientific community to guard against any vulnerability to charges of bias and to do a better job of public education—and on Obama in his role as the nation's persuader in chief.[63]

Ethos refers to a leader's personal integrity and good character, which helps overcome the skepticism of the listener: "Why should I believe what this person says or do what he's urging?" Ethos has another, less personal, but still pertinent meaning: "the characteristic spirit, prevalent tone of sentiment, of a people or community; the genius or guiding spirit of an institution or system."

Over a year into Obama's presidency, public-opinion surveys gave him high marks for his ethos in the first sense—what the pollsters call his leadership qualities and personal characteristics—but during the knock-down, drag-out debate over health care, he took a beating on performance. The president was held responsible for what was widely seen as the dysfuntionality of the federal government over which—as his title says—he presides. Confidence in government fell to an all-time low. An overwhelming majority of Americans—as much as 80 percent in some polls—did not trust the powers that be in Washington to deal competently and effectively with the difficulties besetting the nation.[64]

In tackling climate change, Obama has to find a way of regaining that public trust. To do that, he will need more ammunition for the logos component of his argument. In particular, he will need economic trends that point to improvements in the job and housing markets.

But he will also have to meet the third and most important of Aristotle's standards of persuasive leadership: *pathos*. The word—which is related to "empathy"—means, simply, "feeling."

Aristotle believed that an advocate is more likely to bring his listeners onto his side of an argument if he arouses emotions that make them feel he's on their side—that is, that he knows how they feel, what they want, and what they fear.

On climate change, fear is one emotion that is hard *not* to instill, since the prospects suggested by the facts are scary as hell.

Scaring people is helpful insofar as it gets them to take the matter seriously. But fear alone can be dispiriting—conducive to a sense of helplessness in individuals, unease in the body politic, and enervation in government. Fear combined with anger of the sort that many people currently harbor toward the federal government is toxic for a sitting president, especially when his own party controls Congress.

Franklin Roosevelt intuitively understood that danger and rhetorically ju-jitsued it when he and fellow Democrats swept into office during the depths of the Depression. The most memorable and effective line in his 1933 inaugural address was "The only thing we have to fear is fear itself—nameless, unreasoning, unjustified terror which paralyzes needed efforts to convert retreat into advance." While it took FDR years to turn the nation's fortunes around, he managed early on to inspire and sustain confidence that he could do so; he persuaded the American people that he identified with them, knew what was best for them—that he, and they, were not helpless in the face of some very ugly, fearsome facts.

In America during the Great Depression, the principal cause of fear was the state of the national and global economy. So it is again today. That makes Obama's job in motivating the nation to act on climate change all the more difficult. He must convince a critical mass of the citizenry that his course of action will address the current crisis by saving endangered jobs and creating new ones even as it gives America and the world enough control over what might be called "the future facts" of climate change to head off possible disaster.

WITH THOSE GOALS IN MIND, the administration and its congressional allies are playing down the costs and sacrifices that the management of global warming may entail, and they are playing up the benefits to the economy. Starting with the president, they are selling comprehensive energy legislation as a jobs bill as well as a climate bill.

In his final public appearance in Copenhagen, Obama was speaking primarily to his constituents back home. The issue at hand was how to save the planet from peril decades in the future, but the president concentrated on his determination to return unemployed Americans to work and protect America's independence *now*. He said his "commitment to transform our energy economy at home" had resulted in "historic investments in renewable energy that have already put people back to work"; he was "committed to comprehensive legislation that will create millions of new American jobs, power new industry, and enhance our national security by reducing our dependence on foreign oil."

In the politically fraught months that followed, Obama pounded away on this theme. In his State of the Union address on January 27, 2010, he bundled the prospects of nuclear power plants, advanced batteries, solar panels, breakthroughs in efficient ways to cool and heat homes, new biofuels, clean coal technology, and high-speed intercity rail into a single package with no price tag but a big label advertising it as "jobs for Americans." His overall pitch was to "make clean energy the profitable kind of energy in America." That part of the speech was rewarded with a bipartisan standing ovation. But when he concluded by urging Congress to "save our planet from the ravages of climate change" and "send me legislation that places a market-based cap on carbon pollution," he was greeted with cheers almost exclusively from the Democratic side of the aisle.

In February 2010 Obama's Council of Economic Advisers estimated that the $90 billion of spending under the 2009 Recovery and Reinvestment Act on clean energy would create 720,000 jobs by 2012. Of course, the advisers work for the president himself. Their choice of 2012 as a target year was no coincidence, since that is when he will be running for reelection.

But even with a political discount, the administration's optimism about the economic upside of a transition to a low-carbon economy should not be dismissed. Nicholas Stern, who was chief economist of the World Bank and head of the Government Economic Service in the United Kingdom, published a book in 2009 that made much of the compatibility between sound environmental policy and sustainable economic growth.[65] Jobs directly connected with the promotion of energy efficiency can pay for themselves by cutting costs of energy itself. Installing highly efficient lights and heating systems can pay back initial investments in one to five years. Much the same is true of public investment in weatherization technology, such as energy-efficient windows and doors, and in a "smart" electric grid that allows, for instance, a homeowner with a solar roof to sell unused power back to a utility company.[66] Cities are already saving millions of dollars a year in switching traffic lights from incandescent to light-emitting diodes (LED) systems. Sun Belt cities are installing solar roofs on parking installations, cutting both their fuel bills and auto air-conditioning costs because the cars inside stay cooler.

Another claim that advocates make for green jobs is that they will strengthen American exports. That is harder to prove, because the return on investment is harder to calculate. But there is no question the United States risks falling behind other governments that have already placed big bets on the export of green technology and clean energy.

With a smart grid, big tax incentives, and substantial government subsidies, Germany is ahead of the United States and the rest of the world in exporting solar panels and wind turbines. (The solar power figure is impressive, since Germany has far fewer sunny days than other parts of Europe, not to mention the U.S. Sunbelt.)

China recently moved into first place as the leading player in green energy markets, with investments of nearly $35 billion over the past five years—almost double those of the United States.[67]

PUT ALL THESE FACTORS together and they amount to a strong case that the best defense is a strong offense. The United States can and should match what other countries are doing so that it does not lose out in a new sector of global trade. American public investment in clean energy could have a large multiplier effect, perhaps leading to more jobs than the original investment would suggest.

That argument has the most going for it in the field of research and development. It is often difficult for any single firm to fund the R&D necessary to introduce a new product or service, not to mention a revolutionary breakthrough in infrastructure.

Historically, the best advertisement for public funding of R&D is the Internet, which grew largely out of projects funded by the Defense Advanced Research Projects Agency and the National Science Foundation. Federal funding for research in electrical engineering and computer science (including semiconductor technology) climbed to almost $1 billion by 1995. Similar foresight is required in building ties between the national energy labs and U.S. companies that are eager and able to exploit the potential of solar, wind, and biofuels.

THE COMPETITIVE RATIONALE for green technology assumes that other nations are going to continue to plan for a low-carbon future, and not just vigorously export green technology but also open their own markets to U.S. products. That is not guaranteed to happen, nor is it guaranteed to last forever.

But precisely that uncertainty argues for doing everything possible to keep the current interest in green technology, green R&D, green jobs, and green trade alive. If that interest fades, it will be for one of two reasons or perhaps both: another global economic and financial meltdown, or the collapse of political will to address climate change. Either would be a disaster, and the two together would be even worse. While a global recession (not to mention a depression) would cut back on industrial output and therefore on emissions, it would cost precious time and shatter political will at both the national and international levels for the structural changes necessary to put the world on a low-carbon energy path.

One way to nurture recovery and stave off another downturn is to ensure that the United States, as the most powerful economy in the world, commits itself irrevocably and aggressively to sustaining the new green revolution, with American federal policy as a driver of what happens internationally and the American domestic economy as a beneficiary of the resulting opportunities for export.

SUSTAINABILITY, BY DEFINITION, CAN BE PROVED ONLY over time. But there is a near-term job-saving argument for making it a long-term national goal: jump-starting a green sector of the economy can create new jobs that, by their very nature, cannot be exported the way service and manufacturing jobs have moved offshore in recent decades. Installing renewable-energy and energy-efficiency systems in American homes, offices, universities, and factories; building a high-speed intercity rail infrastructure; manufacturing made-in-the-USA solar panels, high-performance batteries, and new biofuels—these are all activities that would put Americans back to work and the United States back on the offensive in the export market.

Still, among the lingering questions that are sure to figure in the debate over a comprehensive energy and climate bill are these two: Will there be enough new jobs to replace the old jobs that have disappeared or are in jeopardy in energy-intensive industries, such as coal, steel, and heavy manufacturing? And will a massive investment in clean energy produce an economy that is larger and more stable than the old one that crashed in 2008–09?

Economics tells us that fundamental transitions are expensive, and the costs tend to come up-front. The Congressional Budget Office and the Council of Economic Advisers both estimated that the 2009 House climate change bill would cost the average household $80 to $400 a year. Over the next forty years, that would be a total cost of $1 trillion to $5 trillion.*

If a bill emerges from the Senate this year or next, it is likely to have similar goals for cutting CO_2 emissions and therefore very

*The estimates of cost vary widely because so much depends on variables. In the best case, the price signal is clear, and companies respond quickly by implementing off-the-shelf technologies to promote energy efficiency and already existing carbon-free sources. In the worst case, obstacles—such as the inability of start-up solar and wind providers, or even homeowners, to sell their carbon-free power to the electrical grid—drive up costs.

real costs for businesses and average citizens, even if it is technically "deficit-neutral" in its impact on the federal budget. Hence the possibility that the final bill will include a "price collar"—in effect, a ceiling on the cost of converting to a green-clean economy combined with a "floor" on a price for carbon.[68]

THE ECONOMICS OF CLIMATE CHANGE is, in a way, analogous to the science. Just as there is no such thing as a right projection of what the temperature curve looks like, or what will happen when or with what consequences on that curve, there is no clear, indisputable, fact-based bottom line on the cost-benefit analysis of mitigation—and for the same reason: inherent uncertainty about the future.

But, as with the science, uncertainty about the economics is not an excuse for inaction. Quite the contrary: just as prudence requires us to think about the worst that climate change might do in the decades to come, we also have to be prepared for the high end of what it might cost, especially in the near term, to avert the worst.

This much we do know: it will cost *something*. In Washington there is, famously, no free lunch, nor is there ever going to be a free climate bill, even if it is called an energy-independence bill or a green-jobs bill.

So what is the antidote to sticker shock? What is the best way to get politicians and citizens to accept the costs of avoiding climate disaster? In one respect the argument is simple: those costs, whatever they end up being, are *nothing* compared to those of *not* avoiding disaster. Odds are that the cost of business as usual—that is, of relying on fossil fuels like there's no tomorrow—is a very large multiple of the $5 trillion that comprehensive climate legislation might cost. The price of doing nothing, if it leads to global warming at its worst, would be incalculable and irreversible—and not just economic: it would include the degradation and destruction of life, including human life, on a vast scale.

All of which is to say: there *is* a tomorrow, and we must think about it in a new and more serious way.

IN LATE FEBRUARY 2010, Lindsey Graham explained to Thomas L. Friedman of the *New York Times* why he was, at that point, prepared to break ranks with the just-say-no strategy of the Republican party on the issue of global warming and make common cause with his Senate colleagues, Kerry and Lieberman—and with Obama.

"I have been to enough college campuses to know if you are thirty or younger this climate issue is not a debate," said Graham. "It's a value. These young people grew up with recycling and a sensitivity to the environment—and the world will be better off for it. They are not brainwashed. . . . From a Republican point of view, we should buy into it and embrace it and not belittle them. You can have a genuine debate about the science of climate change, but when you say that those who believe it are buying a hoax and are wacky people you are putting at risk your party's future with younger people."[69]

If there were a full record of all that has been said and written about climate change, the word count would probably be at the giga level. Of all those words, Graham's hundred or so contain a hint of what should be, and maybe someday will be, but is not yet the clincher in the case to be made for action on climate change.

Graham was not just preaching to the converted—he was preaching to a preacher. Friedman had been on a crusade about global warming for nearly a decade. Graham's purpose in going on the record in that interview was to convert other Republicans by scaring them: if the GOP makes obscurantism on climate change part of its platform in the midterm elections in 2010 or in the next presidential campaign in 2012, it will be risking its future with young voters.

That was the explicit message. Implicitly, Graham was making a synaptic connection between, on the one hand, the down-and-dirty politics of getting a climate (a/k/a green jobs, a/k/a energy independence) bill to the president's desk before the Cancun summit and, on the other, the metapolitics and ethics of global warming.

Graham is not just a Republican but a conservative. There is no more legitimate and compelling conservative cause than mitigating the risk to the planet and the human race. Graham is also partisan—he would not be a U.S. senator if he were not—and he was making a partisan point: "We Republicans need more young people in our ranks." But his exhortation could, if taken to heart by his party, prompt Republicans and Democrats to compete for young, environmentally conscious voters, thereby having a bipartisan benefit.

If Graham ends up cosponsoring a climate bill, his role in the U.S. legislative melee this year may turn out to be roughly comparable to Obama's role in the diplomatic one in Copenhagen last year: breaking a logjam and bridging a divide. In Obama's case the split was (and remains) between the developed and the developing countries. In Graham's case it would be between Republicans and Democrats. In both cases they may end up having restored just enough momentum to a flagging but vitally important process to keep it moving forward.

But whatever the U.S. government does in the summer and fall of 2010 and whatever the international community does in Cancun at the end of the year will be, at best, a stopgap—a tentative gesture in the direction of what has to be done in the years ahead.

The heavy lifting required to remove, over the next forty years, 15 gigatons of CO_2 emissions a year from the 30 that we are now emitting will demand an exertion of national and international political will vastly more strenuous and effective than anything now in prospect.

Political will is a protean concept. It can mean the ambition—or, in Obama's terms, the audacity—of an individual wanting to acquire political power (which he has demonstrated) and then trying to use it to get his way (which is where he had difficulties in his first year). Then there is the national political will necessary to pull out of a recession, mobilize for war, or rebuild after a war. Finally, there is global political will. There were signs of it during the cold war in the delicate balance of steadfastness and accommodation that was necessary to preserve the nuclear peace. There were signs of it again in late 2008 and early 2009 in the creation of the G-20 and other innovations for dealing with the worldwide financial crisis.

But the ultimate test is whether—as the climate continues to change and the planet continues to warm—there is the collective and coordinated political will of nations to slow that process down. That will depend on nothing less than a revolution in global civics.[70]

THE HIGHER SELFISHNESS

Intellectuals like Kant, Hegel, and Freud despaired over the lag between progress in technology and progress in morality. Yet confronted with the new reality of the atomic bomb, political leaders like Henry Stimson and Harry Truman made the leap from a Clausewitzian world to one in which war between the great powers was no longer an option. So did Joseph Stalin, who was otherwise a moral monster. Because they and their successors knew they bore an unprecedented burden, they acted in new ways—and restrained themselves from acting in old ways. In the face of nuclear war, they expanded their sense of communal responsibility from the traditional one of defending their own nations to the protection of the human race.

That was the logical next step in the ethical and political evolution that Socrates anticipated and advocated in the fifth century BCE. His claim of being a citizen of the world rather than an Athenian or a Greek contributed to the charge of sedition ("refusal to accept the gods of the state") for which he was forced to drink hemlock.[71]

Humanity's single greatest achievement has been, in that sense, Socratic: the expansion of community from the tribe to the city-state to the nation to the world. Now that we have, belatedly, recognized the reality of climate change, we have to take yet a further step in the same direction, toward a sense of community that includes responsibility for the protection of future generations.

In his interview with Friedman, Senator Graham appealed to Republicans' individual *and collective* self-interest in political survival. The idea that "self" can be plural is at the core of Adam Smith's guide to comprehending and influencing human behavior:

> [However] selfish . . . man may be supposed, there are evidently some principles in his nature, which interest him in the fortune of others, and render their happiness necessary to him, though he derives nothing from it except the pleasure

> of seeing it. Of this kind is pity or compassion, the emotion which we feel for the misery of others, when we either see it, or are made to conceive it in a very lively manner. That we often derive sorrow from the sorrow of others, is a matter of fact too obvious to require any instances to prove it; for this sentiment, like all the other original passions of human nature, is by no means confined to the virtuous and humane, though they perhaps may feel it with the most exquisite sensibility. The greatest ruffian, the most hardened violator of the laws of society, is not altogether without it.[72]

Smith is revered as a pioneer in economics, the dismal science that is intimately and often contentiously linked to the politics swirling around climate science. But Smith understood his discipline in the broadest sense to be concerned with ethics: the passage cited above is from a work he titled *The Theory of Moral Sentiments*. The moral as well as political evolution of our civilization has been largely a matter of applying Smith's theory to the practice of governance in increasingly sophisticated, expansive, and interconnected societies and polities.

American political culture is imbued with what might be called the higher selfishness. Alexis de Tocqueville saw the genius of the U.S. system to be its ability to nurture the formation of civic associations on an ad hoc basis so that communal action was driven by what he called "self-interest rightly understood."[73] Starting with the Declaration of Independence's assertion of universal and inalienable rights, American foreign policy has been based on the idea that self-interest understood rightly is self-interest understood globally.

In recent decades, revolutions in transportation and communication have brought into being what amounts to a global society and a system of global governance that is intended, so far with modest but promising success, to advance common interests and manage common threats through a variety of institutions, regulatory structures, and collaborative habits of interstate relations.

We need to keep broadening and deepening that process. But because of climate change, we need, simultaneously, to take on a new challenge: the concept of universal citizenry must not just be spatial in its scope—it must extend to the dimension of time, since those who will live in the future will depend on us who live in the present for the quality of their lives if not for their very existence. Tomorrow's citizens have always been dependent on today's. That was true to an unprecedented degree in the Nuclear Age and is true in the Age of Global Warming.

Previous generations understood that their lives and deeds were chapters in an ongoing saga connecting them to those who had come before and to those who would come after. Edmund Burke saw society and civilization as a "partnership of generations . . . between those who are living, those who are dead, and those who are to be born." He saw members of any one generation as "temporary possessors and life-renters in" society and in the earth; he feared that citizens might become "unmindful of what they have received from their ancestors, or of what is due to their posterity," and therefore run the risk of "leav[ing] to those who come after them a ruin instead of an habitation."[74]

Thomas Jefferson made much the same point when he argued that because "the earth belongs in usufruct [in effect, in trust] to the living," and that while each generation should feel free to determine its own destiny, "no generation can contract debts greater than may be paid during the course of its own existence."[75]

In the twentieth century, Hannah Arendt envisioned, in *The Human Condition,* the "public realm" of a "common world" that would "contain a public space [that] cannot be erected for one generation and planned for the living only; it must transcend the life-span of mortal men."[76]

We are still in the habit of thinking of the continuum from one generation to the next as an accumulation of positive legacies. It has long been a working assumption that children would be at least somewhat better off than their parents: when something good happens to you, you should "pay it forward," rather than pay it back.

According to the legal concept of intergenerational equity, assets do not belong exclusively to those who have accrued them; rather, those resources should, to the extent possible, be administered and preserved for those who will inherit them and will, partly as a consequence of their inheritance, live somewhat better lives than those who came before. We come into this world in debt to our ancestors, and we leave it an incrementally better place, passing along improvements to our descendants, but also leaving them to come up with their own means of fending off or coping with whatever their age throws at them.

That has been the narrative of individual families and of the human family over time.

Global warming alters that narrative in a very basic way. We cannot leave those who come after us to their own devices for dealing with global warming. If we do not get the process of mitigating climate change started now, on our watch, their devices will not be sufficient. We are likely to bequeath them a less habitable—perhaps even uninhabitable—planet, which would be the most negative legacy imaginable.

THERE IS AN EVOCATIVE, somewhat eerie-sounding term from game-theory—"the shadow of the future"—that has found its way into economics. The phrase has been used in reference to the tendency of buyers and sellers to consider transactions with an eye not just to immediate costs and benefits but to longer-term ones as well. In that sense, the future casts a "shadow" over current decisionmaking.

Climate change reverses the meaning of the phrase. Actions or inactions in the present cast a shadow in the other direction, *over* the future. Decisions or indecision today can impose heavy costs on our descendants or, at a minimum, limit the choices they will have. That is why there is an unprecedented need to merge the reality of an international community with the established principle of intergenerational responsibility to form a politically potent and actionable concept of a global intergenerational community. Only then can we widen the definition of our interests beyond the horizon of our own life spans.

Human nature is an obstacle in this regard. The more distant the future, the less we care. Even on fast forward and in its worst-case version, the global warming scenario does not become a true nightmare for several decades.

Yet many of us have children and grandchildren who will then be middle-aged. Parenthood brings with it an intimation of immortality by proxy. It has become common on the climate change lecture circuit for speakers to invoke the names—and, if they are using slides, show pictures—of their own offspring as representatives of the planet's more than 2 billion children who are likely to reap the results, good or bad, of our current leaders' decisions.

James Hansen—one of the earliest scientists to sound the alarm on global warming and one of the most outspoken Paul Reveres of the current climate debate—titled his most recent book *Storms of My Grandchildren.* By personalizing the fate of the next generation with reference to his own family, he is inviting his readers to do the same with theirs. It is a classic example of rhetorical pathos, and as such it is perfectly appropriate (the book you are reading now is dedicated to one author's children, the other's grandchildren). It can even be effective with the right audience, which is often one that does not need a lot of convincing.

But to scale up and project forward the idea of a civic responsibility that is not so much intergenerational as it is *trans*generational (in that it is about our descendants, not our ancestors), we are going to have to *de*personalize the issue of global warming; we are going to have to generalize our empathy and accountability beyond our own great- and great-great-grandchildren whom we will not live to see but whose potential, and eventual, existence is, for many of us, a reason for living and a hedge against our fear of dying.

Nearly thirty years ago, Jonathan Schell, in *The Fate of the Earth,* warned that global thermonuclear war would "doom all future human beings to uncreation." He appealed to his contemporaries to realize they were the first people ever to hold "the life and death of the species in [their] hands," which made them responsible for "guaranteeing the existence of all future generations."[77]

Eighteen years ago, Al Gore began *Earth in the Balance* with the admonition that "we have it in our power to restore the earth's balance before the growing imbalance inflicts its greatest potential damage on our children and grandchildren. Today the human species is the only one with the self-knowledge and the capacity to protect its own future."[78]

Hansen, casting his mind half a millennium into the future to the year 2525, speculates that our "Goldilocks planet" of today (not too cold, not too hot) will be in the throes of "the Venus

Syndrome," which would replicate on earth the scorching heat and the arid, lifeless surface of the next planet nearer to the sun: "I've come to conclude that if we burn all reserves of oil, gas, and coal, there is a substantial chance we will initiate the runaway greenhouse. If we also burn the tar sands and tar shale, I believe the Venus syndrome is a dead certainty."[79]

Those last two words—even preceded by the *if*s—do not belong in any sentence about the future, especially from a scientist. Assertions of certainty about the future are part of the reason that the science of climate change has, from time to time, been open to attack. Besides, one does not have to go to eschatological extremes to establish what Schell calls "communion with the unborn." Even if the race escapes extinction, our successors may not escape the consequences of our policies on climate change, wherever those consequences are on the spectrum from so-so to bad to really, *really* bad.

That recognition alone should be enough for us to see that our descendants will have rights that only we can protect. That gives them a claim on us. If we recognize and discharge that claim, it will be easier to accept the sacrifices that—notwithstanding the emphasis on benefits by advocates of comprehensive climate and energy legislation—the timely mitigation of climate change will almost certainly demand.

It is a shame, though probably one we will have to live with, that the word *sacrifice,* like *tax,* has become almost taboo in American political discourse.

More than shameful, it is ahistorical. Personal as well as communal self-sacrifice has an ancient and noble place in the annals of politics, economics, ethics, and civics. It has long been an accepted obligation that comes with communal self-interest self-governance.

Countless millions have willingly given their lives for the sake of their families, their tribes, their nations, or, in Socrates' case, his world. Yet that extreme sacrifice is not required to preserve the livability of the planet. To defeat—or at least manage—global warming, no one is asking us, or our children, to die at the stake, on the gallows, or on a battlefield.

The demand is much more modest than that: we simply need to incorporate into our decisions at all levels, from private households to the White House and beyond, a determination to be frugal and foresighted in our expenditure of carbon. Frugality is not fatal; it is not even dangerous. It's smart, and it's necessary.

REINFORCEMENT

The appeal to transgenerational responsibility as the best argument for vigorous, and expensive, action on climate change has one big flaw: pretty much everyone pays lip service to the concept, but few let its implications get in the way of thinking about today's utility bills, today's lifestyle, or next November's elections. Investing the concept of transgenerational responsibility with political potency is a tall order. Obama is going to need reinforcements, and not just from fellow politicians.

Just as wars are too important to leave entirely to generals, creating the necessary national and international ethos is too important, and too difficult, to be left entirely to governments. While recent attacks on the IPCC have sown fresh public doubt about climate change, there are countervailing increases in vigorous, assertive support for realism and effective policy from various organizations that make up civil society.

Heads of state and government lead from above, but they are increasingly being either supported or pressured from below, depending on whether they are seen to be leading in the right direction. That mix of support and pressure must come from advocacy groups, nongovernmental organizations, academe, the corporate world, the labor movement, and faith-based communities.

Throughout the world, religion is a force in politics, for good and ill. One area in which it is especially constructive is climate change. Numerous and diverse spiritual leaders are beginning to form a virtual caucus, if not movement, on behalf of urgent and ambitious political action. Muslim clerics, Catholic bishops, Protestant ministers, evangelical preachers, Hindu and Buddhist priests have created a network with links like the one between the Protestant Religious Witness for the Earth to the Catholic Climate Covenant.

Rabbis, particularly social activists, speak of our responsibility today for tomorrow's citizens in the context of *tikkun olam*, a Hebrew phrase often translated as "repairing the world," with the connotation of leaving it a better place than we found it.

The Islamic Foundation for Ecology and Environmental Sciences, made up largely of Indonesians, is trying to rein in deforestation by an appeal to "Islamic environmental norms," such as a precept in the Koran that "greater indeed than the creation of man is the creation of the heavens and the earth."[80]

In 2006 the biologist and ethicist Edward O. Wilson published *The Creation: An Appeal to Save Life on Earth.* It was a plea to Christian fundamentalists to form an alliance on climate change with other religious groups, including those with liberal views on other issues. A number of clerical leaders who tend toward socially conservative positions agree with Wilson that stewardship of God's creation is a cause that transcends traditional politics.

The Rev. Jim Ball, the leader of the Evangelical Environme
Network's Climate Campaign, argues in his forthcoming boo
Global Warming and the Risen Lord, that "all Christians have a contribution to make in overcoming global warming." He has stressed individual responsibility "not only to bear and nurture children, but to nurture their home on earth" at the level of family lifestyle and energy use.

Over 300 senior evangelical leaders, including the Rev. Rick Warren, who gave the invocation at Obama's inauguration, are part of the Evangelical Climate Initiative. They have urged their followers to press for government action. Many Episcopal churches provide weekly bulletins, emphasizing measures from switching to energy-efficient light bulbs to lobbying members of Congress.

Early in 2010 Pope Benedict XVI exhorted Catholics not to "remain indifferent before the problems associated with such realities as climate change, desertification, the deterioration and loss of productivity in vast agricultural areas, the pollution of rivers and aquifers, the loss of biodiversity, the increase of natural catastrophes and the deforestation of equatorial and tropical regions." The U.S. Conference of Catholic Bishops, which played an active part in alerting the public to the danger of nuclear war and the moral dilemma of nuclear deterrence in the 1980s, has called on the heads of the G-8 to help the "poor countries and peoples who have contributed the least" to climate change. "Protecting the poor and the planet are not competing causes; they are moral priorities for all people living in this world."[81]

No religious leader of a major faith has identified himself more closely with meeting the challenge of climate change than has His All Holiness Bartholomew, Ecumenical Patriarch of Orthodox Christianity. He has surprised many of his colleagues and owers with his activism on the environment. He has traveled e world conducting on-site symposia in regions affected by climate change: Greenland, the Black Sea, the Amazon, and the Mississippi. Orthodoxy, he reminds people, does not mean "old-thinking," but "right-thinking"—and the best example of that virtue, he believes, is "steering the earth toward our children."[82]

We know the course we have to set. Now we have to get moving, fast.

ACKNOWLEDGMENTS

WE ARE FORTUNATE IN HAVING been able to draw on the work and advice of scholars from all five Brookings research programs as well as from several of our trustees. That is appropriate since the Institution has designated the subject of climate change and energy an All-Brookings Priority.

Special thanks to colleagues who reviewed our manuscript in draft: Alexis Bataillon, Steve Bennett, Kemal Dervis, Charles Ebinger, Bill Galston, Ted Gayer, Bruce Jones, Richard Kauffman, Ken Lieberthal, Pietro Nivola, Lea Rosenbohm, Les Silverman, John Thornton, and Darrell West (who is the author of the next book in the Focus series, *Brain Gain: Rethinking U.S. Immigration Policy,* to be published in June 2010).

There are also a number of others at Brookings whom we consulted on the general topic and on specific issues: Liaquat Ahamed, Hakan Altinay, Richard Blum, E. J. Dionne, Martin Indyk, Bruce Katz, Teresa Heinz Kerry, Kevin Massy, Warwick McKibbin, Adele Morris, Mark Muro, Urjit Patel, Eswar Prasad, Rob Puentes, Isaac Sorkin, Peter Wilcoxen, Tracy Wolstencroft, and Dan Yergin.

The team of the Brookings-Blum Roundtable deserves recognition for having hosted two meetings on the issue of climate change and global poverty: Abigail Jones, Karen Kornbluh, and Nigel Purvis.

Our gratitude and praise go to the Brookings Institution Press, under the leadership of Bob Faherty, and to Marty Gottron, Janet Walker, and Christopher Kelaher. Melissa Skolfield and her associates in the communications department provided several useful ideas on this project and the Focus series of which it is a part.

Several former Brookings colleagues were helpful in multiple ways before they went into the U.S. government: Jeff Bader, Jason Bordoff, Lael Brainard, Peter Orszag, Carlos Pascual, David Sandalow, and Jim Steinberg.

We have also had indispensable support from Eric Larson, Phil Duffy, and Berrien Moore at Climate Central, who guided us through the scientific complexities of climate change.

Our research assistants, Katie Short and Alex Fife, kept the manuscript moving on a fast-forward schedule.

Our appreciation to Father Alex Karloutsos, Father Mark Arey, and Father John Chryssavgis of the Greek Orthodox Archdiocese of America for the opportunity to speak to present parts of this book during the pan-Orthodox vespers service for the United Nations, and also to His All Holiness Patriarch Bartholomew for speaking at Brookings.

Thanks also to our friends at *Politico,* John Harris and David Mark, who ran op eds which became parts of chapters.

Other colleagues in the public policy community have given us considerable counsel, including specific advice on this manuscript or on material and meetings from which it is drawn: the Reverend Jim Ball, Roger Ballantine, Matt Bennett, Jean Marc Coicaud, Phil DeCola, Elliot Diringer, Paula Dobrianski, Tom Friedman,

Lord Anthony Giddens, Hal Harvey, David Held, Tom Heller, John Ikenberry, Walter Isaacson, Fred Krupp, Claus Leggewie, Michael Levy, Bernhard Lorentz, Alden Meyer, Rakesh Mohan, Jennifer Morgan, Ray Offenheiser, Joe O'Neill, Michael Oppenheimer, Bob Orr, Hermann Ott, Cem Özdemir, Stewart Patrick, Annie Petsonk, John Podesta, Carl Pope, Glenn Prickett, William Reilly, Tim Richards, Mary Robinson, Smita Singh, Mark Suzman, Humayun Tai, Mary Wells, Michael Werz, and Tim Wirth.

We acknowledge with gratitude the help of several people in the Executive Branch of the U.S. government, several of whom were close colleagues before their current public service: Shere Abbott, Joe Aldy, Susan Biniaz, Mike Froman, Jim Kapsis, David Lipton, Peter Ogden, Billy Pizer, Jonathan Pershing, and Todd Stern.

We have had the benefit of private discussions and public Brookings sessions with current and former members of Congress: Lamar Alexander, Dianne Feinstein, Dick Gephardt, Lindsey Graham, John Kerry, Richard Lugar, Nancy Pelosi, and Tom Perriello. Key congressional staff members have also assisted us: Neil Brown, Kathleen Frangione, Frank Lowenstein, and Ken Myers.

Special thanks to Bill Clinton, Al Gore, and colleagues from our own time in government, including Warren Christopher and Madeleine Albright, Sandy Berger, Stuart Eisenstadt, Leon Fuerth, and Dan Tarullo, as well as to Ban Ki Moon, Montek Singh Alluwahlia, Ivo de Boer, Stavros Dimas, Connie Hedegaard, Jairam Ramesh, Meera Shankar, and Rachmat Witoelar.

Each of us has personal debts as well.

Bill is grateful to the University of Nevada, Las Vegas, where he presented parts of this book as a lecture. Particular thanks to Brett Birdsong, Robert Lang, Tom Piechota, Neal Smatresk, and Ron Smith. Thanks also to Becky Boulton, Bill Brown, Matthew

Costello, and Brian Greenspun for helping launch Brookings Mountain West, which includes significant work on clean energy and climate change. Special thanks to Lindy Schumacher and Jeff Wilkins at the Lincy Foundation whose support for Brookings Mountain West was instrumental to this project. Bill also thanks the Taunt Group for their never-ending insights and analyses on politics, diplomacy, and baseball.

In addition, Bill thanks his wife Kristen Suokko, whose own career in nuclear security and on environmental protection has informed Bill's work for over a decade, and who was an irresistible advocate for the project before the co-authors took it on. She pored over the manuscript and fed us throughout a weekend of intense collaboration in Charlottesville during the blizzard of December 2009. Bill also thanks his brother Kary, who helped brainstorm this and a hundred other ideas over the years. Bill derived inspiration from his mother and father, Eva and John, neither of whom lived to see this project completed. But they did both live long enough to see the arrival of the next generation, Annika and Kyri, who share the dedication of this book.

Strobe wants to thank Carole Hall and Amanda Mays for their heroic efforts to manage his life and work during the forced march toward publication; Anne Taylor Fleming for initial encouragement to undertake the project and support throughout; Leonard Schaeffer, Tolliver Besson, Sandra Seitman, Javier Solana, and Concha Gimenez for their patience and encouragement during the final fortnight of work on the book when they were traveling together; and Mindy Barnes, Becky Mooneyham, and Tewelde Seyoum for hospitality, good cheer, and sustenance at key points along the way.

Like Bill, Strobe had sustained help and reinforcement from his family: his brother Kirk, his sisters Marjo and Page, and his brother-in-law and permanent editor Mark Vershbow, as we his sons, Devin and Adrian, both of whom read the book sever times in manuscript and made many useful suggestions. Devin and Lauren's children, Loretta Josephine and Theodore Brooke, have, in their own way, been a source of inspiration—particularly for the final chapter—thereby earning their place alongside Annika and Kyri in the dedication.

Strobe's deepest debt is to his father Bud, who—at least a quarter of a century ago—alerted his children to the phenomenon of global warming and has worked unrelentingly on the issue ever since, and who has, for more than six decades, been Strobe's guide, teacher, and role model, notably including in the way he has incorporated into his civic life an active appreciation of humanity's dependence on and responsibility for the world of nature.

W.J.A. and S.T.
April 2010

NOTES

1. Mark Twain in the *Hartford Courant,* Aug. 24, 1897.

2. See Philip Duffy's blog, "Baby It's Cold Outside" (Climate Central, February 19, 2010) (www.climatecentral.org/breaking/blog/baby_its_cold_outside).

3. James Hansen, *Storms of My Grandchildren: The Truth about the Coming Climate Catastrophe and Our Last Chance to Save Humanity* (New York: Bloomsbury, 2009), p. 71.

4. IPCC, "Summary for Policymakers." In *Climate Change 2007: The Physical Science Basis. Contribution of Working Group I to the Fourth Assessment Report of the Intergovernmental Panel on Climate Change,* edited by S. Solomon, D. Qin, M. Manning, Z. Chen, M. Marquis, K. B. Avery, M. Tignor, and H. L. Miller (Cambridge University Press, 2007) (www.ipcc.ch/pdf/assessment-report/ar4/wg1/ar4-wg1-spm.pdf).

5. The temperature rise from "preindustrial" levels is based on surveys of tens of thousands of measurements made at weather stations around the world starting in the late nineteenth century. For further explanation, see Philip Duffy, "Taking the Earth's Temperature" (www.climatecentral.org/breaking/blog/taking_the earths_temperature).

6. For a clear explanation about the difficulties with two degrees, see John Wibhey's "The Meaning of the G-8's '2 Degrees' Goal; Adequate? Realistic? Too Vague? A Distraction? Maybe" (Yale Forum on Climate Change and the Media, August 4, 2009) (www.yaleclimatemediaforum.org/2009/08/g8s-2-degrees-goal).

. Hansen, *Storms of My Grandchildren,* pp. 250–58. Also see mas Karl, Jerry Melillo, Thomas Peterson, and Susan Hassol, eds., *obal Climate Change Impacts in the United States* (New York: Cambridge University Press, 2009), pp. 27–39 (http://downloads.globalchange.gov/usimpacts/pdfs/climate-impacts-report.pdf).

8. CNA Corporation, *National Security and the Threat of Climate Change* (Alexandria, Va.: 2007).

9. While there is debate about how much CO_2 and other human-induced greenhouse gases are contributing to warming, most recent findings—including from America's top climate scientist—suggest that a 400 ppm target is a modest estimate of what is needed. That figure—400 ppm of CO_2—would lead to a total concentration of about 450 ppm of CO_2 and other anthropogenic greenhouse gases (often referred to as CO_2 equivalent, or CO_2e). The IPCC has provided a number of different scenarios, depending on the two biggest variables: how much CO_2 contributes to warming, and how the hydrological cycle works. In one scenario, a doubling of CO_2 and other greenhouse gases above prehistoric levels would lead to temperature change of somewhere between 3.6°F–5°F (2°C–3°C). The 400 ppm CO_2 (450 ppm CO_2e) figure has been widely adopted as the goal among IPCC scientists. However, James Hansen, perhaps the leading climate change scientist in the U.S. government, and a team of colleagues have recently argued that 350 ppm CO_2 is the appropriate target. See James Hansen, Makiko Sato, Pushker Kharecha, David Beerling, Valerie Masson-Delmotte, Mark Pagani, Maureen Raymo, Dana L. Royer, and James C. Zachos, "Target Atmospheric CO_2: Where Should Humanity Aim?" (www.columbia.edu/~jeh1/2008/TargetCO2_20080407.pdf).

10. Cited in Nigel Purvis and William Antholis, "Case for Climate Protection Authority," Politico, January 27, 2009 (dyn.politico.com/printstory.cfm?uuid=14D3AB9B-18FE-70B2-A879566FAF5DF6B8). Richard Gephardt, former House Democratic Leader, served in the U.S. House of Representatives for twenty-eight years as a representative from Missouri.

11. See, for example, Obama's commencement address, Notre Dame, May 17, 2008: "You have a different deal. Your class has come of age at a moment of great consequence for our nation and for the world—a rare inflection point in history where the size and scope of the challenges before us require that we remake our world to renew its promise; that we align our deepest values and commitments to the demands of a new age.

It is a privilege and a responsibility afforded to few generations—a a task that you're now called to fulfill." The next day, Vice Presider Joseph Biden, at a commencement ceremony in Wake Forest, Illinois, said, "It's a different world out there than it has been any time in the last millennia. But we have an opportunity to make it beautiful, because it is in motion. We have an opportunity to change it. But absent our leadership, it will continue to careen down the path we're going now. And that could be terrible. That, folks, is an inflection point."

12. Andrew S. Grove, *Only the Paranoid Survive* (New York: Doubleday, 1996).

13. Smithsonian Institution, National Museum of Natural History, Department of Botany, "Scientists at the Smithsonian's National Museum of Natural History Find Global Warming to be Major Factor in Early Blossoming Flowers in Washington" (www.mnh.si.edu/highlight/spring00/spring00_feature.html). Also see Paul K. Strobe, "Implications of Climate Change for North American Wood Warblers (Parulidae)," *Global Change Biology* 9 (August 2003) (http://courses.nres.uiuc.edu/nres456/Strode%20warbler.pdf.) For information on the increasingly warm winters, see National Oceanic and Atmospheric Administration, National Climatic Data Center, "State of the Climate Global Analysis," February 2010. The report notes: "The combined global land and ocean average surface temperature for December 2009–February 2010 was the fifth warmest on record for the season, 0.57°C (1.03°F) above the 20th century average of 12.1°C (53.8°F)."

14. Immanuel Kant, *Kant's Prolegomena to Any Future Metaphysics,* translated into English by Dr. Paul Carus (Chicago: The Open Court Pub. Co., 1902).

15. G. W. F. Hegel, *The Philosophy of History* (New York: Cosimo, 2007), pp. 20–21.

16. Sigmund Freud, *Civilization and Its Discontents,* translated and edited by James Strachey (New York: Norton, 2005).

17. Michael Walzer, *Just and Unjust Wars: A Moral Argument with Historical Illustrations* (New York: Basic Books, 1977), pp. 269–83.

18. Jonathan Schell, *The Fate of the Earth* (New York: Knopf, 1983).

19. One reviewer, Theodore H. Draper, was scathing. "Utopian obscurantism," he fumed, "a travesty of thinking about nuclear war," "the most depressing and defeatist cure-all." See his "How Not to Think about Nuclear War," *New York Review of Books* (July 15, 1982).

20. U.S. Conference of Catholic Bishops, *The Challenge of Peace: God's Promise and Our Response: A Pastoral Letter on War and Peace by the National Conference of Catholic Bishops* (Washington: May 3, 1983).

21. Henry L. Stimson and McGeorge Bundy, *On Active Service in Peace and War* (New York: Harper, 1948), pp. 635–36.

22. Richard Rhodes, *The Making of the Atomic Bomb* (New York: Simon & Schuster, 1986), pp. 690–91.

23. David Holloway, *Stalin and the Bomb: The Soviet Union and Atomic Energy, 1939–1956* (Yale University Press, 1994), pp. 335–45. Also see Thomas Schelling, "An Astonishing Sixty Years: The Legacy of Hiroshima," The Sveriges Riksbank Prize in Economic Sciences in Memory of Alfred Nobel 2005, December 8, 2005. In his lecture, Schelling noted: "Yet the Soviets spent great amounts of money developing non-nuclear capabilities in Europe, especially aircraft capable of delivering conventional bombs. This expensive capability would have been utterly useless in the event of any war that was bound to become nuclear. It reflects a tacit Soviet acknowledgement that both sides might be capable of non-nuclear war and that both sides had an interest, an interest worth a lot of money, in keeping war non-nuclear—keeping it non-nuclear by having the capability of fighting a non-nuclear war."

24. Zhu Minguan, "The Evolution of China's Nuclear Nonproliferation Policy," *Nonproliferation Review* 4 (Winter 1997).

25. Robert Maynard Hutchins, "Atomic Force: Its Meaning for Mankind," University of Chicago Round-Table of the Air, NBC broadcast, August 12, 1945, transcript p. 12.

26. On the eve of passage in 1990, the EPA predicted that the sulfur dioxide cap-and-trade program would cost $6 billion annually. By 2002 the Congressional Budget Office estimated that the actual cost was somewhere between $1.1 and $1.8 billion, just 20 to 30 percent of the amount forecast.

27. Jeremy Leggett, *Carbon War: Global Warming and the End of the Oil Era* (New York: Penguin Books, 1999), p. 72.

28. Clinton interview with Talbott, March 24, 2010.

29. Jody Warrick and Peter Baker, "Clinton Details Global Warming Plan," *Washington Post,* October 23, 1997, p. A1.

30. Ibid.

31. The incentive for clean energy projects in developing countries was called the Clean Development Mechanism. While it has been popular

among environmentalists, it also has been controversial. See Mark Scha-piro, "Conning the Climate: Inside the Carbon-Trading Shell Game," *Harper's Magazine,* February 2010.

32. Cullen Murphy and Todd S. Purdham, "Farewell to All That: An Oral History of the Bush White House," *Vanity Fair* (February 2009).

33. William Antholis, "The Good, The Bad, and The Ugly: EU-US Cooperation on Climate Change," conference paper for "The Great Transformation: Climate Change as Cultural Change," Essen, Germany, June 10, 2009 (www.brookings.edu/~/media/Files/rc/speeches/2009/0610_climate_antholis/0610_climate_antholis.pdf.)

34. As reported by Rick Piltz, a senior associate in that endeavor, the annual report had been "drafted with input from dozens of federal scientists and reviewed and vetted and revised and vetted some more." As it was about to be sent to Congress, a Bush appointee—Phil Cooney, who had been a lobbyist at the American Petroleum Institute—edited the document to introduce uncertainty where scientists felt there to be none. As Piltz would later say, "The political motivation of it was obvious." Murphy and Purdham, "Farewell to All That."

35. Our Brookings colleagues in Metropolitan Policy and Governance Studies have done pathbreaking work on chronicling and analyzing these trends, as well as pointing to better policy design. See Marilyn A. Brown, Frank Southworth, and Andrea Sarzynski, "Shrinking the Carbon Footprint of Metropolitan America," Policy Brief (Brookings, Metropolitan Studies, May 2008) (www.brookings.edu/~/media/Files/rc/reports/2008/05_carbon_footprint_sarzynski/carbonfootprint_report.pdf); Barry G. Rabe and Christopher P. Borick, "The Climate of Opinion: State Views on Climate Change and Policy Options," Issues in Governance Studies 19 (Brookings, September 2008) (www.brookings.edu/~/media/Files/rc/papers/2008/09_climate_rabe_borick/09_climate_rabe_borick.pdf); and Barry G. Rabe and Christopher P. Borick, "The Climate of Belief: American Public Opinion on Climate Change," Issues in Governance Studies 31 (Brookings, January 2010) (www.brookings.edu/~/media/Files/rc/papers/2010/01_climate_rabe_borick/01_climate_rabe_borick.pdf).

36. William J. Baumol, Robert W. Crandall, Robert W. Hahn, Paul L. Joskow, Robert E. Litan, and Richard L. Schmalensee, "Regulating Emissions of Greenhouse Gases under Section 202(a) of the Clean Air Act," *Amicus Curiae* Brief 06-01 to the U.S. Supreme Court, October 2006.

see also Warwick J. McKibbin and Peter J. Wilcoxen, "The Role of Economics in Climate Change Policy," *Journal of Economic Perspectives* 16 (2002), p. 107; and William D. Nordhaus and Joseph Boyer, *Warming the World: Economic Models Of Global Warming* (MIT Press, 2000), pp. 121–44.

37. Judy Keen, "Bush Prepares for Standoff with Blair at G8," *USA Today,* July 5, 2005 (www.usatoday.com/news/world/2005-07-05-g8-bush-blair_x.htm).

38. One of the authors of this book had proposed a similar kind of summit between rich and poor nations on climate change several months earlier. See Todd Stern and William Antholis, "Creating an E-8," *American Interest* (January-February 2007).

39. See, for example, the Friends of the Earth press release of December 19, 2009: "Brokenhagen—Climate Summit Ends in Failure" (www.foe.co.uk/resource/press_releases/copenhagen_19122009.html).

40. John Broder, "Poor and Emerging States Stall Climate Negotiations," *New York Times,* December 16, 2009.

41. For more on the incident, see Phil Duffy, "Mistakes Happen" (www.climatecentral.org/breaking/blog/mistakes happen).

42. Moisés Naím, "Minilateralism: The Magic Number to Get Real International Action," *Foreign Policy* (July-August 2009).

43. Michael Oldstone, *Viruses, Plagues, and History* (Oxford University Press, 1998), p. 27; and Richard Preston, "Demon in the Freezer," *New Yorker* (July 12, 1999), pp. 44–61. The authors are indebted to the late Professors Ernest May and Erez Manela of Harvard for assistance on this subject.

44. Paul Kennedy, *The Parliament of Man* (New York: Random House, 2006), pp. 206–08, 238, and 285–88. Kennedy took his title from a poem by Alfred Lord Tennyson, "Locksley Hall," that Harry Truman cherished as an expression of his ideal for the United Nations. The key stanza reads:

> Till the war-drum throbbed no longer, and the battle-flags were furl'd
> In the Parliament of Man, the Federation of the World.
> There the common sense of most shall hold a fretful realm in awe,
> And the kindly earth shall slumber, lapt in universal law.

45. William Antholis, "The Good, The Bad, and The Ugly: EU-US Cooperation on Climate Change." For useful comparisons between U.S. and European efforts to regulate automobile fuel efficiency, see Pietro

Nivola, "The Long and Winding Road: Automotive Fuel Economy and American Politics," Policy Brief (Brookings, Governance Studies, February 25, 2009, www.brookings.edu/~/media/Files/rc/papers/2009/0225_cafe_nivola/0225_cafe_nivola.pdf).

46. The U.S. Department of Energy's Energy Information Agency forecasts Indian emissions to increase to just under 2 gigatons by 2020. But that is wildly low compared with even the Indian government's own assessment. An experts' committee established by the government of India estimated that in the year 2030 CO_2 emissions would likely reach between 4 and 5.5 gigatons, depending on the fuel mix. See *Integrated Energy Policy: Report of the Experts' Committee* (New Delhi: Government of India, 2006), p. 50. McKinsey's number is 5.7 gigatons CO_2e, with about 0.5 gigatons coming from noncarbon greenhouse gas emissions. See Rajat Gupta, Shirish Sankhe, and Sahana Sarma, *Environmental and Energy Sustainability: An Approach for India* (McKinsey & Company, August 2009).

47. N. C. Saxena, "Climate Change and Food Security in India," in *Climate Change, Perspectives from India,* ch. 3 (United Nations Development Programme, India, November 2009) (www.undp.org.in/content/pub/ClimateChange/UNDP_Climate_Change.pdf). The report cites: "The existing problems of poor farmers, if not addressed in time, will become more acute due to global warming induced climate change. The prediction so far suggests an upward trend in mean monthly temperature and average rainfall. However, the prediction indicates a downward trend in the number of wet days in a year. The impact of climate change would be seen in terms of increased sub-regional variations and more extreme rain events. In a country that gets rain for less than 100 hours in a year (a year has 8,760 hours), this would be disastrous." Additional news reports include Roger Harrabin, "How Climate Change Hits India's Poor," BBC News, February 1, 2007; and Saibal Dasgupta, "Warmer Tibet Can See Brahmaputra Flood Assam," *Times of India,* February 3, 2007. More generally see Abigail Jones, Vinca LaFleur, and Nigel Purvis, "Double Jeopardy: What the Climate Crisis Means for the Poor," in *Climate Change and Global Poverty,* edited by Lael Brainard, Abigail Jones, and Nigel Purvis (Brookings, 2009).

48. Information Office of the State Council of the People's Republic of China, *China's Policies and Actions for Addressing Climate Change* (Beijing: 2008) (www.gov.cn/english/2008-10/29/content_1134544.htm).

49. Warwick McKibbin, "Environmental Consequences of Rising Energy Use in China," paper prepared for the Asian Economic Policy Review conference, October 22, 2005.

50. Environmental Information System, Ministry of Environment and Forests, Government of India, *State of Environment Report: India – 2009* (New Delhi: July 20, 2009), p. 78.

51. The 2030 projections are included both in Urjit Patel, "Decarbonization Strategies: How Much, How, Where and Who Pays for $\Delta \leq 2°C$?" Conference Paper, Fortieth Annual Conference of the Gujarat Economic Association, and in Gupta, Sankhe, and Sarma, *Environmental and Energy Sustainability.*

52. Several such plans have been laid out. See Vijay Joshi and Urjit Patel, "India and Climate Change Mitigation," in *The Economics and Politics of Climate Change,* edited by Dieter Helm and Cameron Hepburn (Oxford University Press, 2009). See also Gupta, Sankhe, and Sarma, *Environmental and Energy Sustainability;* and Patel, "Decarbonization Strategies."

53. Rashme Sehgal, "India to Set Up 21 Nuclear Projects in Five Years—Daily," *BBC Monitoring South Asia,* October 14, 2008. The article notes: "India is kick-starting its nuclear facilities by setting up 21 nuclear power projects based on three different technologies. These include the setting up of six French reactors of 1,600 MW, four Russian reactors of 1,000 MW and four American reactors of 1,500 MW within the next five years." Also, "India Signs Contract with French Firm to Build Atomic Plants," *BBC Monitoring South Asia,* February 4, 2009, cites, "Ending 34 years of nuclear isolation, India Wednesday [February 4] signed its first commercial contract to build atomic power plants here with French company Areva. Initially, Areva is expected to supply two European Pressurised Reactors (EPR) of 1650 MW each which are likely to be set up at Jaitapur in the western state of Maharashtra." Los Angeles currently has 7,500 MW of capacity. Los Angeles Department of Water and Power (www.ladwp.com/ladwp/cms/ladwp001557.jsp).

54. Kenneth Lieberthal and David Sandalow, *Overcoming Obstacles to U.S.-China Cooperation on Climate Change,* (Washington: Brookings, John L Thornton China Center, 2008), p. 33.

55. Keith Bradsher, "Nuclear Power Expansion in China Stirs Concerns," *New York Times,* December 15, 2009.

56. Elizabeth Economy, "Testimony before Senate Foreign Relations Committee, United States Senate, June 4, 2009."

57. "Politburo Set on Carbon Emissions Target; Party Leaders Ru Out Further Debate on Goal Promised at Copenhagen Meeting," *South China Morning Post,* February 16, 2010.

58. For a version of the GARE proposal, see Todd Stern and William Antholis, "A Changing Climate: The Road Ahead for the United States," *Washington Quarterly* 31 (Winter 2007–08), pp. 175–88. See also Annie Petsonk, "Testimony before the Subcommittee on Energy and Air Quality," Committe on Energy and Commerce, March 27, 2007, and Nigel Purvis, "Trading Approaches on Climate: The Case for Climate Protection Authority," *Resources* (Summer 2008).

59. U.S. House of Representatives, "American Clean Energy and Security Act of 2009," 111 Cong. 1 sess., HR 2454, Title VII, Part C, Section 728, International Emissions Allowances, p. 774.

60. In 2008 the G-20 countries (including all EU-27) were responsible for 25.4 gigatons of the world's total 30.4 gigatons of CO_2. See U.S. Energy Information Administration, *International Energy Statistics, 2009* (2009).

61. Obama was at least the third president to invoke Socrates but the first to do so as a candidate. In January 20, 1961, John F. Kennedy addressed "my fellow citizens of the world" in his inaugural address. On June 17, 1982, at the UN General Assembly, Ronald Reagan said, "I speak today as both a citizen of the United States and of the world." On July 24, 2008—at the height of his campaign for the White House—Obama began a speech to a crowd of over 200,000 people in the Tiergarten, an expansive park in the center of Berlin, "Tonight, I speak to you not as a candidate for president, but as a citizen—a proud citizen of the United States, and a fellow citizen of the world." He was pilloried by conservative columnists and bloggers.

62. Aristotle, *Rhetoric,* translated by W. Rhys Roberts (New York: Random House, 1954).

63. See Elizabeth Kolbert, "Up in the Air," *The New Yorker,* April 14, 2010.

64. See CNN's survey released on January 12, 2010, which asked whether Obama had "the personality and leadership qualities a President should have." Sixty-four percent of the respondents said yes, 35 percent said no. An NBC/*Wall Street Journal* survey released that same week asked, "How confident are you that Barack Obama has the right set of personal characteristics to be president of the United States?" Fifty-one

percent answered "Extremely/quite." On the issue of performance, that same NBC/*Wall Street Journal* poll asked: "Compared to what you expected when Barack Obama took office a year ago, do you feel that he has done better than you expected, worse than you expected, or just about as you expected?" Only 15 percent said better than expected, while 30 percent said worse. By contrast, Bill Clinton's performance was rated "worse than expected" by only 15 percent at the comparable point in his presidency. See William Galston, "In Government America Must Trust," an op-ed in the *Financial Times,* March 3, 2010.

65. Nicholas Stern, *The Global Deal: Climate Change and the Creation of a New Era of Progress and Prosperity* (New York: PublicAffairs, 2009).

66. Mark Muro and Robert Puentes, "Growth through Innovation: It Will Be Metropolitan-Led" (Brookings, November 2009).

67. Pew Charitable Trusts, "Who's Winning the Clean Energy Race? Growth, Competition and Opportunity in the World's Largest Economies" (Washington: March 25, 2010) (www.eenews.net/public/25/14924/features/documents/2010/03/25/document_cw_03.pdf).

68. On this issue, too, Brookings scholars are in the forefront of generating analysis and proposals. See Warwick McKibbin, Adele Morris, and Peter Wilcoxen, "A Copenhagen Collar: Achieving Comparable Effort through Carbon Price Agreements," paper presented at the Brookings Blum Roundtable in August 2009, in Aspen, Colo.

69. Thomas L. Friedman, "How the G.O.P. Goes Green," *New York Times,* February 28, 2010.

70. Hakan Altinay, a Nonresident Senior Fellow at Brookings, is developing a project on global civics that will provide a curriculum on the subject for universities around the world. See www.brookings.edu/papers/2010/03_global_civics_altinay.aspx?rssid=altinayh.

71. Plutarch, in *Moralia,* vol. 3, "Of Banishment," writes Socrates "said . . . he was not an Athenian or a Greek, but a citizen of the world (just as a man calls himself a citizen of Rhodes or Corinth), because he did not enclose himself within the limits of Sunium, Taenarum, or the Ceraunian mountains."

72. Adam Smith, *The Theory of Moral Sentiments* (London: Henry G. Bohn, 1853), p. 3.

73. Alexis de Tocqueville, *Democracy in America,* trans. Henry Reeve (New York: J. & H. G. Langley, 1840), p. 122.

74. Right Hon. Edmund Burke, M.P, "Reflections on the Revolution in France and on the Proceedings in Certain Societies in London Relative

to that Event in a Letter Intended to have been Sent to a Gentleman in Paris" (London: Rivingtons, 1868), p. 110.

75. Merrill Peterson, ed, *Jefferson Writings: Autobiography, Notes on the State of Virginia, Public and Private Papers, Addresses, Letters* (New York: Literary Classics of the United States, 1984), pp. 959–60.

76. Hannah Arendt, *The Human Condition* (University of Chicago Press, 1958), pp. 52–55.

77. Jonathan Schell, *The Fate of the Earth* (New York: Alfred A. Knopf, 1982), pp. 177, 173.

78. Al Gore, *Earth in the Balance: Ecology and the Human Spirit* (New York: Houghton Mifflin, 2000), p. XII.

79. James Hansen, *Storms of My Grandchildren: The Truth about the Coming Climate Catastrophe and Our Last Chance to Save Humanity* (New York: Bloomsbury, 2009), p. 236.

80. Hadrat Mirzā, Nāsir Ahmad, and Malik Ghulām Farīd (eds.), *The Holy Qur'an* (London Mosque, 1981), ch. 40:58, p. 1016.

81. Letter from National Conference of Catholic Bishops to the Leaders of the G-8 Nations (Washington: June 22, 2009). See also the letter to members of Congress sent on behalf of the U.S. Conference of Catholic Bishops by Bishop Howard Hubbard and Ken Hackett, president of Catholic Relief Services, June 22, 2009.

82. His All Holiness Ecumenical Patriarch Bartholomew, *Encountering the Mystery* (New York: Doubleday, 2008), p. 101. See also his remarks at Brookings, "Saving the Soul of the Planet," November 4, 2009.

INDEX